LIFE ON THE EDGE

Studies in Social Ecology and Environmental History

General Editors: RAMACHANDRA GUHA and
MADHAV GADGIL

Other Books in the Series

DAVID ARNOLD AND RAMACHANDRA GUHA (EDS)
Nature, Culture, Imperialism: Essays on the Environmental History of South Asia (Oxford India Paperbacks)

AMITA BAVISKAR
In the Belly of the River: Tribal Conflicts Over Development in the Narmada Valley (Oxford India Paperbacks)

BARBARA BROWER
Sherpa of Khumbu: People, Livestock and Landscape (Oxford India Paperbacks)

MADHAV GADGIL AND RAMACHANDRA GUHA
This Fissured Land: An Ecological History of India (Oxford India Paperbacks)

KRISHNA GHIMIRE
Forest or Farm?: The Politics of Poverty and Land Hunger in Nepal

RICHARD GROVE, VINITA DAMODARAN AND SATPAL SANGWAN (EDS)
Nature and the Orient: The Environmental History of South and Southeast Asia

MAHESH RANGARAJAN
Fencing the Forest: Conservation and Ecological Change in India's Central Provinces, 1860–1914 (Oxford India Paperbacks)

CHETAN SINGH
Natural Premises: Ecology and Peasant Life in the Western Himalaya

VASANT SABERWAL
Pastoral Politics: Shepherds, Bureaucrats, and Conservation in the Western Himalaya

AJAY SKARIA
Hybrid Histories: Forests, Frontiers and Wildness in Western India

PURNENDU S. KAVOORI
Pastoralism in Expansion: The Transhuming Herders of Western Rasjasthan

K. SIVARAMAKRISHNAN
Modern Forests: Statemaking and Environmental Change in Colonial Eastern India

Life on the Edge

Sustaining Agriculture and
Community Resources in
Fragile Environments

N. S. Jodha

OXFORD
UNIVERSITY PRESS

OXFORD
UNIVERSITY PRESS

YMCA Library Building, Jai Singh Road, New Delhi 110001

Oxford University Press is a department of the University of Oxford. It furthers
the University's objective of excellence in research, scholarship, and education
by publishing worldwide in

Oxford   New York

Athens Auckland Bangkok Bogota Buenos Aires
Cape Town Chennai Dar es Salaam Delhi Florence Hong Kong Istanbul
Karachi Kolkata, Kuala Lumpur Madrid Melbourne Mexico City Mumbai
Nairobi Paris Sao Paolo Singapore Taipei Tokyo Toronto Warsaw

with associated companies in Berlin Ibadan
Oxford is a registered trade mark of Oxford University Press
in the UK and in certain other countries

Published in India
By Oxford University Press, New Delhi

ISBN 019 565134 0 ✓

Typeset in 10 on 12 Galliard
By Wordsmiths, Delhi-110034
Printed at Saurabh Print-o-Pack, NOIDA, UP
Published by Manzar Khan, Oxford University Press
YMCA Library Building, Jai Singh Road, New Delhi 110 001

To the people and communities
of dry tropical areas and mountains
whose wisdom and ways
in adapting to fragile environments
induced and enhanced
my involvement
in promoting the issues and messages
communicated through this volume

Foreword

Poverty, environmental degradation, and the inability of communities to live with one another have dominated societal concerns in developing countries for many years now. Even though much of the academic literature that has been developed on these problems would give an impression to the contrary, theoretical and empirical findings in recent years suggest strongly that none of the three is a prior cause of the other two, but rather, that each influences the others and is in turn influenced by them.

For example, a few economists have studied the interface of poverty and the local natural-resource base in poor countries. The ingredients of their work have been around for some time; what is perhaps new is the way they have been put together. The work has involved a fusion of theoretical modelling with empirical findings drawn from a number of disciplines, most notably, anthropology, demography, ecology, economics, and the nutritional and political sciences. I do not suppose their work amounts to a theory, it is more like a new perspective. One of its aims has been to offer an explanation for the fact that in recent decades large groups of people in various parts of the world have been unable to lift themselves out of acute poverty even when their national economies have, on average, enjoyed economic growth. At the grandest level, the aim has been to create a set of analytical tools to account for the fact that, even though the world economy as a whole has enjoyed economic growth over the past fifty years or so, large masses of people in certain regions have remained in poverty. Economic growth has not 'trickled down' consistently to the poorest, nor have the poorest been inevitably 'pulled up' by it. Cross-section and intertemporal statistics on national aggregates, such as per capita income, life expectancy, and literacy, can mislead because they involve the use of weak lenses for peering at the social world. If persons are to count, persons need to be the starting and returning points of analysis. This requires the use of a strong lens. The new perspective adopts this modelling device.

I should stress that several particular models have been constructed to develop the new perspective. We still are nowhere near an overarching model, of the kind economists are used to in the theory of general competitive equilibrium.[1] Some models have as their ingredients large inequalities in asset ownership in poor countries and the non-convexities which prevail at the level of the individual person in transforming nutrition intake into nutritional status and thereby, labour productivity. Others are based on the fragility of interpersonal relationships in the face of an expanding labour market and an underdeveloped set of credit and insurance markets. Yet others are built on possible links between fertility behaviour and free-riding on local common property resources once communitarian sharing arrangements break down.[2] The models differ in their ingredients. What they have in common, however, is a structure which is becoming increasingly familiar from the theory of locally interacting systems.[3] To put it in contemporary terminology, the new perspective on poverty and the local natural-resource base sees the social world as self-organizing into an *in*homogeneous whole, so that even while parts grow, chunks get left behind; some even shrink. To put it colloquially, these models account for locally confined 'vicious circles', or what Myrdal (1944) called 'cumulative causation'.

Needless to say, it is not assumed in the new perspective that poor, illiterate people, when subjected to such 'forces' of positive feedback, do not try to find pathways by which to cope. The models assume that people do the best they can in the circumstances they face. What the models do is to identify conditions in which this is not enough to lift communities out of the mire. Turner and Ali (1996), for example, have shown that, in the face of population pressure in Bangladesh, small landholders have periodically adopted new ways of doing things so as to intensify agricultural production. However, the authors have shown too that this has resulted only in an imperceptible improvement in average well-being and a worsening of the ownership of land, the latter owing to the prevalence of distress sales of land. This is the kind of finding which the new perspective anticipated, and was designed to meet.

[1]In this, the literature I am alluding to resembles much contemporary economic theory.

[2]For accounts of such models, see Dasgupta (1993, 1997). See also Homer-Dixon et al. (1993) on battles for resources.

[3]Brock and Durlauf 1999 is a fine technical account of that structure.

What suggestions are there for improving matters in a world subjected to such positive feedback processes? If in earlier days social scientists looked for policies to shape social outcomes for the better, their focus today is more on the character of institutions within which decisions are made by the myriad of parties within and without society. But if policies which read well often come to naught in dysfunctional institutions, the study of institutions on their own is not sufficient: good policies cannot be plucked from air. There is mutual influence here, and the task before the social scientist is to study it.

Economists have traditionally been much engaged in the study of markets, political scientists in the study of the state, and anthropologists and sociologists in the study of interpersonal networks. It is only in recent years that each group has begun to peer into the others' publications to see if they can better understand the links that connect their particular objects of interest. The interplay of market and non-market institutions is rich in possibilities and offers the social scientist a potentially most exciting research agenda. The new perspective I have been referring to has been directed at studying this interplay.

Consider, for example, the experience people in poor countries have had with structural adjustment programmes, which involved reductions in the plethora of economic distortions that had been introduced by governments over decades. Many have criticized the way the programmes have been implemented. They have pointed to the additional hardship many of the poor have experienced in their wake. But it is possible to argue that structural adjustments, facilitating as they did, the growth of markets, were necessary. And it has been so argued by proponents of the programmes. The new perspective suggests that *both* proponents and opponents of the programmes may have been right. Growth of markets benefits many, but it can simultaneously make vulnerable people face additional economic hardship and so increase the incidence and intensity of poverty and destitution in an economy.

How and why might this happen? There are a number of pathways by which it can happen. Here is one:

Long-term relationships in rural communities in poor countries are typically sustained by the practice of social norms, for example, norms of reciprocity. This is not the place to elaborate upon the way social norms should technically be viewed (as self-enforcing strategies).

The point about social norms which bears stressing, however, is that they can be practised only among people who expect to encounter one another repeatedly in similar situations.

Consider then a group of 'far-sighted' people who know one another and who prepare to interact indefinitely with one another. By a far-sighted person I mean someone who applies a low rate to discount future costs and benefits of alternative courses of action. Assume as well that the parties in question are not separately mobile (although they could be collectively mobile, as in the case of nomadic societies); otherwise the chance of future encounters with one another would be low and people (being far-sighted!) would discount heavily the future benefits of current cooperation.

The basic idea is this: if people are far-sighted and are not separately mobile, a credible threat by all that they would impose sufficiently stiff sanctions on anyone who broke the agreement would deter everyone from breaking it. But the threat of sanctions would cease to have potency if opportunistic behaviour were to become personally more profitable. This can happen during a process in which formal markets grow nearby and uncorrelated migration accompanies the process. As opportunities outside the village improve, those with lesser ties (e.g. young men) are more likely to take advantage of them and make a break with those customary obligations that are enshrined in prevailing social norms. Those with greater attachments would perceive this, and so infer that the expected benefits from complying with agreements are now lower. Either way, norms of reciprocity could be expected to break down, making certain groups of people (e.g. women, the old, and the very young) worse off. This is a case where improved institutional performance elsewhere (e.g. growth of markets in the economy at large) has an adverse effect on the functioning of a local, non-market institution. To the extent local common property natural resources are made vulnerable by the breakdown of communitarian control mechanisms, structural adjustment programmes would have been expected to be unfriendly also to the environment and so, to those who are directly dependent on them for their livelihood. This is because when the market value of a resource base increases, there is especial additional pressure on the base if people have relatively free access to it. Structural adjustment programmes devoid of safety nets for those who are vulnerable to the erosion of communitarian practices are defective. They can also be damaging to the natural environment unless the structure of property rights, be they private or communitarian, is simultaneously made more secure. We should not have expected matters to have been otherwise.

The above example indicates that an essential ingredient in any model of institutions is a specification of property rights on goods and services. The latter include not only such aggregates as land and cattle and water, but also such finely honed objects as fruit and berries and branches of trees in semi-arid villages. Our understanding of the roles of property rights and their enforcement in the performance of institutions is now a great deal better than what it was even twenty years ago.

Recent happenings in Seattle and the response of those who regard free trade as being good for everyone can be understood, even if very partially, when viewed through the new perspective. Public discussions on the appropriate role of the World Trade Organization (WTO) are now routinely conducted in terms of an alleged battle between multinational companies and hapless governments in poor countries. But the poor in poor countries are not the same as the governments who rule over them. To be sure, increased international trade has benefited many and arbitrary restrictions on trade have harmed also many. But freeing trade in the presence of incompletely specified and only partially enforced property rights can be predicted to hurt segments of the population and has been known to hurt them. The economics of the 'second-best' is today capable of identifying the kinds of people who would be expected to get hurt when trade expansion occurs in the absence of appropriate safety nets or compensations.

Consider, for example, the ecological pathways by which deforestation in the uplands of a watershed inflict damage on people in the lowlands. It pays to study the pathways in terms of the assignment of property rights. The common law in many poor countries, if we are permitted to use this expression in a universal context, in principle recognizes pollutees' rights. So it is the timber merchant who, in principle, would have to pay compensation to the farmers for the right to inflict the damage that goes with deforestation. However, even if the law sees the matter in this light, there is a gulf between the 'written' law and the enforcement of law. When the cause of damage is hundreds of miles away, when the timber concession has been awarded to public land by government, and when the victims are a scattered group of impoverished farmers, the issue of a negotiated outcome does not usually arise. If the timber merchant is not required to compensate the downland farmers, the private cost of logging is less than its social cost. So we would expect excessive deforestation of the uplands. We would also expect that resource-based goods would be underpriced in the market. The less roundabout the production of the final good, the greater would this underpricing be, in percentage terms. Put another way, the lower the value that is added to the resource in the course of production, the larger is the extent of this underpricing of the final product. Thus, when property rights are not enforced in countries which export primary products, there is an implicit subsidy on the exports, possibly on a massive scale. Moreover, the subsidy is paid not by the general public via taxation, but by some of the most disadvantaged members of society: the sharecropper, the small landholder or tenant farmer, the fisherman. The subsidy is hidden from public scrutiny, that is why it is not acknowledged officially. But it is there. It is real. We should be in a position to estimate such subsidies. As of now, we have very few official estimates. Since expansion of trade could be expected to increase the commercial value of such primary products as timber, the link between the gains and losses from international trade and the enforcement of property rights should be made to rear its head when discussions on the role of WTO are undertaken. Modern economic analysis can identify scenarios where the gains would be less than the losses. In such circumstances increased trade without a concomitant improvement in the enforcement of property rights would be harmful to a nation in the aggregate. Even for WTO governance is at the heart of the matter, not trade.

None of the theorizing that is involved in these examples could have progressed in the way it did had there not been strong empirical evidence in support of it. Common forms of non-market institutions are the plethora of communitarian arrangements that exist throughout the world for the management of local common property resources. Communitarian systems have protected people in many places from the 'tragedy of the commons'. In poor countries the local commons include grazing lands, threshing grounds, swidden fallows, inland and coastal fisheries, rivers and canals, woodlands, forests, village tanks, and ponds. Are they extensive? As a proportion of total assets, their presence ranges widely across ecological zones. In India the local commons are most prominent in arid regions, mountain regions, and unirrigated areas, they are least prominent in humid regions and river valleys. There is a rationale for this, based on the human desire to reduce risks. An almost immediate empirical corollary is that income inequalities are less where common property resources are more prominent. Aggregate income is a different matter though, and it is the mountain regions and unirrigated areas which are the poorest. This needs to be borne in mind when public policy is devised. As would be expected, dependence on common property resources even within dry regions declines with increasing wealth across households.

Even though economic theorists in the decade of the 1970s had charted arguments which show that communication systems of management can in principle work (see, for example, Dasgupta and Heal 1979), I do not believe many of them had any idea how important such systems could be in the lives of the rural poor. I personally had no sense of the extent of their presence. In a justly celebrated article, Dr Narpat Jodha reported evidence from over eighty villages in twenty-one dry districts in India to conclude that, among poor families, the proportion of income based directly on the commons is for the most part in the range 15–25 per cent (Jodha 1986). Such evidence did not, of course, prove that the local commons in Jodha's sample were well managed, but it showed that rural households would have strong incentives to devise arrangements whereby they would be well managed.

It is hard to describe today how vital this paper was for the development of the new perspective I have tried to sketch here. The article offered a peg on which one could hang a justification for the study of the interplay of market and non-market institutions in poor countries. Subsequent publications by Jodha himself and by such other eminent social scientists as Elinor Ostrom have made the local commons among the most intensively studied objects of enquiry in environmental and resource economics.

How do poor people live on a day-to-day basis? How much of their daily activity is governed directly by the market and how much by non-market arrangements? We need to know such things if only because we need to know why and where such partitioning occurs. No economics of poverty can get off the ground if it is not prompted by such questions. Even on them Jodha has been a pioneer investigator. He has studied people's daily behaviour in the semi-arid plains of central India and the mountain regions of northern

India and Nepal and has given us a feel of how the ecological landscape shapes human lives. The volume represents applied social science at its best. The essays have depth, they have breadth, and they have style.

PARTHA DASGUPTA

REFERENCES

Brock, W.A. and S.N. Durlauf, 1999. 'Interactions-Based Models', Social Systems Research Institute, University of Wisconsin, Madison. Forthcoming, J.J. Heckman and E. Leamer (eds), *Handbook of Econometrics* 5 (Amsterdam).

Dasgupta, P., 1993. *An Inquiry into Well-Being and Destitution,* Oxford: Clarendon Press.

———, 1997. 'Nutritional Status, the Capacity for Work and Poverty Traps', *Journal of Econometrics* 77(1): 5–38.

———, 2000. 'Reproductive Externalities and Fertility Behaviour', *European Economic Review* (Papers and Proceedings).

——— and G. Heal, 1979. *Economic Theory and Exhaustible Resources,* Cambridge: Cambridge University Press.

Homer-Dixon, T., J. Boutwell and G. Rathjens, 1993. 'Environmental Change and Violent Conflict', *Scientific American* 268(2): 16–23.

Jodha, N.S., 1986. 'Common Property Resources and the Rural Poor', *Economic and Political Weekly* 21: 1169–81.

Myrdal, G., 1944. *An American Dilemma: The Negro Problem and Modern Democracy,* New York: Harper & Row.

Turner, B.L. and A.M.S. Ali, 1996. 'Induced Intensification: Agricultural Change in Bangladesh with Implications for Malthus and Boserup', *Proceedings of the National Academy of Sciences,* 93, 14984–91.

Preface

This volume contains eleven essays having clear operational orientation, selected out of the many which I have written over the last
twenty-five years. Though written in different contexts and at different
points of time as well as covering diverse issues and addressed to
diverse audiences, the essays have a strong degree of continuum in
terms of: (i) their central thematic focus, namely the changing
dynamics of nature–society interactions, reflected through the
decisions and actions of the communities which survive on and
contribute to the degradation (and at times to the regeneration) of
the natural resources; (ii) area focus covering fragile resource zones,
namely dry tropical plains and mountains; and (iii) development of
my own professional understanding of emerging issues relating to
the marginal areas and their inhabitants, and their projection and
advocacy to induce the mainstream's concern for them. These issues
relate to the changing status and usage pattern of natural resources
in fragile areas and their underlying processes, and finally indicating
the possibilities of arresting the negative trends characterizing the
above changes.

The volume is a synthesis and a product of my sustained work for
nearly two decades in dry tropical areas of India and over a decade in
the Hindu Kush–Himalayan region. Despite their biophysical
differences, these regions do share several common elements, such as
fragility- and marginality-induced nature and invisibility of their
problems as well as the opportunities and their neglect by the
mainstream policy-makers; people's adaptation strategies to high-
risk–low-productivity environment through folk technologies and
various formal institutional arrangements; gradual decline of
traditional resource management systems and the emerging indicators
of unsustainability of present patterns of resource use. Based primarily
on micro-level empirical evidence from the above regions, most of
the essays are focused on specific problems with environmental and
socio-economic dimensions.

Though the subjects covered by this volume range from drought management to social dimensions of biodiversity conservation, or from sustainability of fragile zone agriculture to driving forces behind environmental resource degradation, including the crisis of rural common property resources, the issues focused on reveal specific strong linkages between the essays, which in turn once again demonstrate the need for an integrated approach to understand and deal with the environmental and socio-economic problems of fragile areas. Viewed differently, the sequence of essays also reveals the gradual evolution of focus, ideas and approaches towards the challenges and opportunities in fragile resource areas.

The primary audience of these essays has largely been the people or agencies dealing with the problems and specific issues covered by them. Consequently, the abridged or enlarged versions of most of these essays have been disseminated in various forms such as discussion papers, issue papers, journal articles, chapters in books of readings, training documents, and policy briefs. This volume would, it is hoped, facilitate ready access to these materials at one place.

The contents and conclusions of these essays have been influenced both by the feedback from the audience or target groups (including government departments, NGOs, donors, students or trainees) to whom the specific messages were initially addressed, and the critical comments (both positive and negative) from several social and natural scientists. I had opportunity to work with the latter at the Central Arid Zone Research Institute (CAZRI), Jodhpur; Agro-Economic Research Centre (AERC), Vallabh Vidyanagar (Gujarat); Indian Agricultural Research Institute, New Delhi; International Crops Research Institute for Semi-Arid Tropics (ICRISAT), Hyderabad; International Institute of Tropical Agriculture (IITA), Tanzania/Nigeria; Environment Department at the World Bank, Washington, DC; and International Centre for Integrated Mountain Development (ICIMOD), Kathmandu. Their contributions are acknowledged.

Several persons have helped in shaping and promoting the book. Ramachandra Guha and Bela Malik through their persistent interest and encouragement significantly contributed in seeing it in its present shape. The secretarial assistance provided by Rajendra Shah and Samjhana Thapa at ICIMOD needs special mention. Finally, I would like to sincerely acknowledge the support and need-based professional input provided by ICIMOD in the finalization of this volume.

N.S. JODHA

Acknowledgements

The essays in this volume (some of them with a few editorial changes, including changed titles) were originally published in various journals or thematic books. I sincerely thank the publishers for their permission to reprint these essays. Giving full credit to them as the original publishers, I thankfully acknowledge the permission/support received from:

Economic and Political Weekly (Mumbai) for
'Famine and Famine Policies: Some Empirical Evidence', *EPW* 10(41) (1975), included as Chapter 1 titled 'Famine Policies: How Apt?' in this volume;
'Sustainable Agriculture in Fragile Resource Zones: Technological Imperatives', *EPW* (Quarterly Review of Agriculture) 26(13) (1991), included as Chapter 3 titled 'Sustainable Agriculture in Fragile Resource Zones';
'Drought Management: The Farmers' Strategies and Their Policy Implications', *EPW* (Quarterly Review of Agriculture) 26(39) (1991), included as Chapter 2 titled 'Farmers' Strategy and Its Relevance for Drought Management';
'Common Property Resources: Contribution and Crisis', *EPW* (Quarterly Review of Agriculture) 25(26) (1990), included as Chapter 5 titled 'Common Property Resources in Crisis';
'Common Property Resources, and the Environmental Context: Role of Biophysical versus Social Stress', *EPW* 30(51) (1995), included as Chapter 7 titled 'Biophysical and Social Stresses on Common Property Resources';
'Poverty and Environmental Resource Degradation', *EPW* 33(36–7) (1998) included as Chapter 10 titled 'Poverty and Environmental Resource Degradation: Alternative Explanation and Possible Solutions' in this volume.

Springer-Verlag & Co., KG, Berlin for a chapter entitled 'Enhancing Food Security in a Warmer and More Crowded World: Factors and Processes in Fragile Zones' from their volume *Climate Change*

xviii • *Acknowledgements*

and World Food Security, edited by Thomas E. Downing, pp. 381–419 (1995), included as Chapter 4 as 'Food Security in Fragile Zones'.

Cambridge University Press, UK for a chapter entitled 'Reviving the Social System–Ecosystem Links in Himalayas' from their publication *Linking Social and Ecological Systems*, edited by Fikret Berkes and Carl Floke, pp. 285–310 (1998), included as Chapter 9.

Annals of Arid Zone (Arid Zone Research Association of India, CAZRI, Jodhpur), for 'Grazing Lands and Biomass Management in Western Rajasthan: Micro Level Evidence', 34(2) (1995), included as Chapter 6.

Contents

Illustrations and Tables

MAPS

CHARTS

TABLES

BOXES

ANNEXURES

Introduction: The Challenges and Opportunities in Fragile Resource Areas

I

The essays presented in this volume focus on different dimensions of the dynamics of nature–society interactions in fragile resource zones. The latter includes dry tropics (covering arid and semi-arid tropical areas or dry areas of India), and mountain (including hills) areas of Hindu Kush–Himalayan region), particularly the areas falling in India, Nepal, Pakistan and China. Dry tropics and mountains, despite their biophysical differences, do share several common elements, such as invisibility of their problems induced by their fragility and marginality and their neglect by the mainstream policy-makers; people's adaptation strategies to high-risk–low-productivity environment, through folk technologies and formal/informal institutional arrangements; gradual decline of traditional systems of resource use and emerging indicators of unsustainability of present patterns of resource use. The essays deal with micro-level empirical evidence as well as macro-level synthesis of observations and evidence assembled on micro-level situations in dry areas and mountain regions. Binding the essays together is a strong degree of continuum in terms of their central thematic focus, namely the changing dynamics of nature–society interactions, reflected through the decisions and actions of the communities which survive on and contribute to the degradation (as well as to the regeneration) of their natural resources; and the development of my own professional understanding of emerging issues relating to the marginal areas and their inhabitants, and their projection and advocacy to induce the mainstream's concern for them. These issues relate to the changing status and usage pattern of natural resources in fragile areas and their underlying processes, and the possibilities of arresting the negative trends characterizing these changes.

The primary audience of these essays has largely been the people or agencies dealing with the problems and specific issues covered by them. Apart from feedback from the audience or target groups (e.g. government departments, NGOs, students or trainees, etc.) to whom the specific messages were addressed, the critical comments (both positive and negative) from several social scientists and natural scientists have helped in shaping the contents and conclusions of these essays.

The essays are grouped under three sub-themes. Chapters 1–4 look at nature–society interactions or social responses to environmental imperatives, by focusing on the agricultural uses of natural resources in fragile resource zones and their implications in terms of risk and uncertainty faced by the farmer in the agriculturally marginal environments.

The next thematic cluster concerns the farmers' coping strategy against environmental constraints (e.g. drought) which is not evolved in isolation from total farming systems. Under the integrated coping strategy, sustenance and livelihood security are closely linked to complementarity of farming activities based on common property resources (CPRs) and private property resources (PPRs). Besides highlighting the role of CPRs as support lands for crop and animal farming, the focus of this thematic cluster moves towards uncultivated community resources and the linkage between environmental resource stability and economic security. CPRs are used as a window to project the social aspects of natural resource management. CPRs constitute an important component of community-based natural resources, which perform several ecological, economic and social functions.

Synthesis of empirical evidence (data and observations) and the understanding of resource degrading processes affecting natural resources (in dry tropics and mountains) constitutes the primary focus of essays under the third thematic cluster. The essays under this group are focused on the close examination of changing socio-economic circumstances including: breakdown of traditional social systems–ecosystem linkages; population changes including disintegration of collective stakes in community resources, decline of social capital and group action; side-effects of public interventions and market penetration, etc., which seem to have imparted primacy to short-term socio-economic concerns over the long-term ecological/biophysical concerns in society's approach to its natural resource base. The essays under this group, besides questioning the mainstream view on the subject, also indicate possible approaches to help revive community involvement in the management of natural resources in fragile resource zones.

II

The first essay in the collection, based on very intensive study of famine and famine relief in the arid western Rajasthan, was written as a rejoinder to an essay entitled 'Needed a New Famine Policy' in the *Economic and Political Weekly* in 1995 by Morris D. Morris. The immediate provocation was a long and detailed argument on famines and famine policies in India, which sounded logical but did not match with the field realities I had intensively studied and documented during the drought years in the early 1960s in the villages of western Rajasthan—one of the most famine- (or drought-) prone areas of India. The essay, with detailed data at farm-plot level, household level and village level, documents how villagers perceive drought and its impact and how they adjust to it through a sequence of activities such as curtailing consumption and cost of asset (livestock) maintenance, collective sharing and migration, etc. The essay quantified all such variables as indicators of increasing stress as time passed after the failure of rain. The essay pleads for understanding and strengthening of the farmer's adjustment mechanisms to drought and scarcity.

By the late 1980s, various research studies, and public debate on the subject due to high incidence of drought years during the 1970s and 1980s, helped in sensitizing the public agencies and thereby helped in reorienting some components of drought management policies of the government. In order to demonstrate the change in policies and programmes and its impact, the Department of Agriculture and Cooperation, Government of India (in collaboration with the Indian Institute of Management, Ahmedabad), organized a National Workshop on the subject, involving relevant government departments, NGOs and independent researchers. This served as an important opportunity to me to synthesize the evidence and insights on the farmer's short- and long-term measures to manage drought and its impact. The immediate concern, however, was to compare elements of farmers' strategies with the drought management measures promoted by the state and central governments. The key message of the second essay is that the farmer (including the pastoralist) does not manage drought in isolation from the overall farming system. The key elements of the farmers' coping strategy are moisture security, biomass stability, flexibility and diversification of activities, resource/product recycling, collective sharing arrangements and managing of a sort of 'asset depletion replenishment cycle' over a period of bad and good years.

This coping strategy is under severe strain. Public measures have in most cases substituted people's initiatives with several negative consequences. There is a need for identifying and harnessing the complementarities between the two.

With the widening of the search area to explore and understand nature–society interaction from adaptation to periodic droughts and scarcity to total farming systems covering all land-based activities including cropping, livestock rearing, agro-forestry and rural commons, etc., there was a rapid emergence of two areas of policy concern and public discourse. One related to global warming-induced climatic change, while the other focused on sustainability. The understanding on the farmer's adjustments to drought and weather risk served as an input in a larger debate on potential adaptations to greenhouse warming-led environmental changes in dry tropics and mountain areas, with an altogether different audience. Besides, the weather-risk related work was more closely integrated into agricultural research planning, involving input from other groups such as meteorologists, plant breeders, agronomists, hydrologists, etc.

By the late 1980s sustainability became a major focus of development discourse even in developing countries. Both in the dry tropics and mountains, which I studied as fragile resource zones, the nature of responses to the imperatives of limited accessibility, fragility, marginality, diversity, and niche opportunities determined the processes conducive or unfavourable to sustainability of agriculture in these regions. The key message of the third essay is to understand the resource specificities of fragile areas and their imperatives, then see the match/mismatch between the imperatives and attributes of agricultural programmes and activities. The larger the mismatch, the greater the chances of unsustainability, as shown by indicators relating to resource health, production flows and management options.

Food security issues in fragile zones are ignored in the global environment debate, on account of the latter's skewed perspectives. The fourth essay discusses the biophysical context of food security in the fragile resource zones, by relating the range of production and consumption options to the imperatives of resource specificities, followed by a discussion of the socio-economic contexts of food systems and their changes following the developments such as population growth, indiscriminate resource use intensification, disregard of demand management as well as unequal exchange relations with the external world. The potential impacts of global warming on the food systems of fragile zones are indicated in terms of the likely accentuation of the prevailing negative trends resulting from disregard of imperatives of the resource specificities of fragile zones.

While working in villages of East Africa during the severe drought of 1980, it was seen that despite complete failure of crops, and several predictions of scarcity-induced deaths, the villagers in the severely affected districts of Tanzania managed to survive through diverse supplies from uncultivated community lands, which are seldom recorded. CPRs is one such source falling outside the formal statistical records. Similar was thought to be the case in India, where the bulk of the rural poor living below the official poverty line would probably not exist today but for the support from village CPRs. The fifth essay summarizes and analyses the results of my most referred study conducted over a period of four years covering nearly ninety villages in eight states of India's arid/semi-arid tropical zone.

The sixth essay, based on household- and village-level data from selected districts of Rajasthan, collected at different points of time for over twenty years, helps capture the dynamics of natural resource use, including complementarities between CPR and PPR, between annual and perennial and between crop and livestock farming in arid parts of Rajasthan. Its central message relates to the crucial role of biomass availability and stability in livestock-dominated mixed farming systems in the high-risk production environment of the desert.

The seventh essay deals with the changing primacy of ecological versus social considerations in influencing the situation of CPRs in India's dry regions. Of late, despite little change in agro-ecological circumstances, i.e. degree of marginality, there has been a decline in the area, productivity and upkeep of CPRs even in the more marginal areas. This is a result of rapid changes in socio-economic circumstances, which compared to their biophysical environment, act as a more effective force influencing the status of CPRs. The essay describes these changes and their consequences at regional, community and farm-household levels. The essay points out that unless the biophysical constraints in the dry areas are substantially reduced, the deliberate marginalization of CPRs would mean reduced range of locally managed and used options for the people to withstand the environmental stresses. This is more so for the rural poor who continue to significantly depend on CPRs for sustenance.

The essays under Part III build upon the evidence and insights acquired through the foregoing essays. Ever since the Earth Summit in 1992, biodiversity conservation has not only become a part of major global agenda, but has helped in cornering disproportionately large sums of donor money for this purpose through Global Environment Facility (GEF), etc. Paradoxically, the global discourse on the subject is focused on biodiversity in wilderness, protected areas, forests, etc., bestowing recognition and responsibilities on governments and international agencies (with their national counterparts) rather than on people as the key custodians of global biodiversity. People are in fact treated as biodiversity damaging agents and excluded as part of the operating biodiversity systems. Increased money for environmental concerns following the increased noise by environmental lobbies has led to reduced international funding (including through the World Bank) for agriculture (both research and development). This change was partly attributed to the perceived conflict between environmental protection and agricultural development. The eighth essay argues for space for people-managed biodiversity. People in many developing countries manage biodiversity in their petty crop fields and home gardens by way of diversified cropping and in their support lands/ wastelands, i.e. CPRs, through protecting and using diverse wild species. Both crop fields and CPRs constitute parts of the agricultural landscape. In comparison to the global efforts for biodiversity conservation through protected areas and wilderness, the costs of a community-centred approach would be negligible, whereas once cumulated over vast areas of traditional agriculture, their contribution to global biodiversity may be huge. Their contribution in terms of providing wild relatives for improving crop varieties, as any crop scientist would aver, cannot be valued in nominal terms. Besides, recognition and support to people's systems would mean recognition and learning from indigenous knowledge systems, which no other approach to biodiversity conservation can offer.

The ninth essay, in keeping with the central thrust of the volume, pleads for natural resource management systems which can meet social needs without being insensitive to protection and regeneration requirements of the community's natural resources. It looks at the social system–ecological system links in terms of a dynamic process of two-way adaptation. Accordingly, the communities, with functional knowledge of their natural resource base (NRB) adapt their needs and usage systems to the limitations and potential of their NRB on the one hand and try to amend/adapt the NRB to meet their changing demands on the other. In open and externally linked areas the NRB usage systems are governed less by ecological concerns and more by man-made circumstances. The latter are often not only insensitive to the limitation of NRB, but are not often under the control of local communities. Consequently, they are more resource-extractive, unsustainable and less conducive to a two-way adaptation. Several illustrations to support the argument, presented in the essay, might inspire innovative workers to evolve practical approaches to make communities more sensitive to their environmental problems.

The unquestioned acceptance of the argument on Poverty–Environmental Resource Degradation Link (P-ERD) by the mainstream discourse not only blocks the avenues for alternative explanations, but offers effective excuses for policy-makers to stick with the 'impossibles' (i.e. inability to prevent resource degradation because they are unable to eradicate poverty) and not to review the side-effects of their intervention (e.g. centralization and top-down measures), which might have contributed more than P-ERD towards natural resource degradation in the fragile resource zones.

Under the traditional systems protection, regeneration and regulated use of the community's resources was largely a result of the community's collective stake in its natural resources. The side-effects of external interventions, as well as institutional and technological changes have adversely affected the community's collective arrangements and approach to their NRB. The essay identifies the elements of traditional systems and pleads for identification of their present-day functional substitutes.

The concluding chapter, dealing with the globalization process and the future of fragile ecosystems, is a case by itself as it does not have the well-organized quantified evidence and analysis characterizing other essays. Based on extrapolations of the past understanding of the fragile area situation and some recent observations, it throws up the issues and ideas which may induce deep thinking, research, and practical steps to minimize negative repercussions of rapid globalization for environment and society in the fragile resource zones.

III

Irrespective of the motivation and circumstances leading to the work underlying different essays in the volume, the latter have been influenced by and to some extent influenced the work on the focused themes by different researchers. To acknowledge this fact and also to give a broad idea of the range of other work that addressed the themes and issues covered by this volume, a quick, surface inventory of the relevant literature is presented below. This will also help put the contents and message of the essays and the volume in general, in the overall context of academic and policy discourse on the subject, i.e. dynamics of nature–society interactions in fragile ecosystems.

Drought, Impact and Adjustment: Literature focusing on drought and famine, their impact and people's adjustments, covers most parts of the dry lands of India and also different drought years. For instance, N.S. Jodha (1969, 1975), L.P. Bharara (1982, 1993); Kavoori (1999) and R.J. Fisher (1997 based on work during 1983–4), focus on western Rajasthan. K.M. Chaudhari and M.T. Bapat (1975) cover drought-affected areas in Gujarat and Rajasthan. M.A. Chen (1991) also covers areas in Gujarat. T.K. Meti (1976), V.V. Borkar and M.V. Nadkarni (1975) and G.K. Karanth (1991) focus on the impact of drought in the rural areas of Karnataka. N.P. Prasad and P. Venkatarao (1996) focus on the drought-affected area of Andhra Pradesh. S.M. Pandey and J.N. Upadhyay (1976) deal with the dry areas in Haryana. M.C. Swaminathan et al. (1969) report on scarcity and hunger in drought-affected areas of Bihar. V. Subramanian (1975) provides a graphic account of the impact of the 1972–3 drought in Maharashtra. The *Economic and Political Weekly* (1985) and J. Pradhan (1993) focus on Kalahandi area in Orissa. The results of these and many other studies on drought, its impact and people's adjustments, in several parts of the country have been synthesized by N.S. Jodha (1978), M. Nadkarni (1985), B. Agarwal (1990), M. Dasgupta (1987), GOI (1989), Jean Drèze (1990) and J. Bandyopadhyay (1987).

While the descriptive account of the impact of droughts and people's adjustments during different years in different areas continues to get the attention of researchers, the discourse on the subject has been taken to different levels and in different directions. Accordingly the whole problem of weather-induced instability and scarcity was projected as a problem of risk and uncertainty. This involved researchers from social sciences and natural sciences as well as several field-level functionaries from NGOs and government agencies (GOI 1989). Visualization of drought-induced crises as a risk problem is analysed by several studies such as H.P. Binswanger (1986), S. Gadgil et al. (1988), Hazell et al. (1986), T.S. Walker and N.S. Jodha (1986) and J.R. Anderson (1991).

Approaches to drought-related risk management included focus on credit and prevention of depletion of family assets (Jodha 1981); crop insurance (Dandekar 1976; Hazell et al. 1986; Parthasarathy and Shameem 1998); provision of alternative employment opportunities such as through employment guarantee scheme (Reynolds and Sundar 1977; Dandekar and Sathe 1980) and its subsequent incarnations culminating in the Jawahar Rozgar Yojana.

Another dimension of drought-related work addressed the issues of food security. Beginning with the advocacy for converting the relief programmes into productive asset-building investment via food-for-work (GOI 1989), the focus on food security during drought years converged on food security through PDS for the poor in general during all years. S. Indrakanth (1997), K.R. Venugopal (1992), R. Radhakrishna and C.H. Rao (1994) and Dev and Suryanarayana (1997) have reflected on different aspects of PDS. T.E. Downing et al. (1996) and Jean Dreze (1990) explore different aspects of food security in the changing demographic, environmental and political contexts.

The creation of durable defence mechanisms against the impact of drought through specifically focused agricultural research and technology constituted another stream of work on the subject, addressed by both natural scientists and social scientists (Anderson 1991; Gadgil et al. 1988; Bidinger and Johansen 1988; Hazell et al. 1986; Walker and Ryan 1990; Anderson and Jodha 1994).

Sustainability of Agriculture in Fragile Areas: The discourse on the short-term instability of agriculture manifested by drought-induced scarcities and measures to protect against them, gradually converged into concern for long-term sustainability of agriculture in fragile areas and even in the 'green revolution' areas. S.A. Vosti and T. Reardon (1997), R. Chand and T. Haque (1997), N.S. Jodha et al. (1992), N.S. Jodha (1995a, 1996a) and M.L. Whitaker et al. (1991) address different aspects of sustainability and development of agriculture. The process of thinking in this direction was accelerated by the global discourse following the events such as the Report of the World Commission on Environment and Development in 1987, the Earth Summit in Rio in 1992, the scare created by projected scenarios on climate change and its regional impacts, and the rapid degradation of environmental resources (Agarwal 1986; Jodha 1986a) in the fragile areas in particular. These developments also enhanced the flow of funds for sustainability-related work, which again induced increased engagement of both social scientists and natural scientists in this field of work. This also encouraged an integrated approach to all land-based activities (e.g. crop, forest, pasture, etc.). Accordingly, agricultural scientists emphasized both crop-centred and resource-centred research and related work. In the process the geographic focus of work also shifted from individual countries to the international context. The need for learning from the indigenous system and involvement of local communities in resource management and technology development also emerged as an important aspect of work. Some of the studies focusing on these aspects include M.L. Parry et al. (1988), J. Farrington and S.B. Mathema (1991), C.H. Hanumantha Rao (1994), F.R. Bidinger and C. Johansen (1988), J. Pretty (1995), M.H. Rao (1992), Vyas and Ratna Reddy (1998), S. Krishna (1996) and P. Dasgupta (1997).

Common Property Resources: An important feature of discussions on agricultural sustainability in fragile regions is the better recognition of linkages between sustainable performance of agriculture and sustainability of its environmental resources, including common property resources, as the latter provide several visible and invisible inputs to crop- and livestock-based mixed farming in these regions (Jodha 1992). Consequently, at policy and programme levels both advocacy and action are increasingly emphasizing the sustainability issues on more integrated basis. This is amply reflected by the content of more operationally focused and widely read publications, such as the *State of India's Environment: Citizens Reports* for different years by the Centre for Science and Environment; the *Honey Bee* (a quarterly magazine covering indigenous practices and farmers' innovations by SRISTI, an NGO located at the Indian Institute of Management, Ahmedabad); *Development Alternatives,* Delhi, *Down to Earth* (a monthly publication from the Centre for Science and Environment, New Delhi); *Waste Land News* from SPWD (Society for Promotion of Wasteland Development), New Delhi; *HIMAL, South Asia* from Kathmandu; and the publications from ASTRAL (Centre for Appropriate Science and Technology for Rural Areas) at the Indian Institute of Science, Bangalore, which focus on combining science with practical realities.

Common property resources (CPRs) did not receive the attention of researchers and policy-makers until the early 1980s. A substantive work covering more than 80 villages in 21 districts of 8 dry tropical states of the country (Jodha 1986) was followed by a large number of studies, many of them supported by the Ford Foundation. The initial studies covered different geographic areas as well as different categories of CPRs such as community pasture, forest, water, watersheds, etc. They focused on the current status of and changes in CPRs, including their area, management systems and productivity; and their contributions to village economy (specially the poor) and environment.

Ecosystem–Social System Links: A number of these initiatives on forest and watershed have largely been supported by external donors, which shows their persistent disregard by investment planners in the country. Besides, due to the continued domination of top-down approach and inadequate understanding of grass-root level realities, they have had only mixed success. For instance, shortage of plantable lands controlled by communities, shortage of plantable material, forest departments' focus on timber rather than fuelwood, and non-involvement of communities, etc. contributed to the limited impact of social forestry programme, as reported by ODA (1989), SIDA (1988), J.M. Kerr et al. (1996) and J.E.M. Arnold and W. Stewart (1991).

The issues of natural resource degradation and associated poverty and food insecurity on the one hand and the decline of traditional CPR management systems (without creating any effective substitutes), on the other, were brought to the forefront by field-based CPR studies. They have helped in building a major agenda for socio-political discourse in the country (Guha 1997). Focusing on different target agencies dealing with policy-programme interventions, the evidence and understanding on CPRs have been synthesized by different studies. Some of them include: B. Agarwal (1990), J.E.M. Arnold and W. Stewart (1991), C. Singh (1986), M. Nadkarni et al. (1989), Chambers et al. (1989b), M. Gadgil and R. Guha (1992, 1995), M. Poffenberger and B. McGean (1996), S. Roy (1996), Kumar et al. (1992) and Arora (1994). This synthesis and advocacy of CPRs represents some form of up-scaling of the work on CPR, which was initially focused on the study of the status of specific categories of CPRs in selected areas. This has already helped in sensitizing the government agencies, NGOs, communities and academia to the problems and prospects of CPR management as a part of sustainable development especially in mountains and dry areas (GOI, 1998).

The practical context to the emerging concerns for CPRs and

their participatory management is provided by initiatives such as the programmes on (a) social forestry (World Bank 1989; SIDA 1988; Blair 1986); (b) joint forest management (Poffenberger and McGean 1996; ICIMOD 1995); (c) use of watershed approach to the rehabilitation of degraded commons and crop lands (GOI 1998; Shah 1998) and (d) the emerging concern and advocacy of social dimensions of biodiversity conservation, including protection of community interests through focus on issues relating to intellectual property rights to communities in natural resources, protected and regenerated by them for ages (Singh 1995; Kothari and Anuradha 1997; Shiva 1996). Community gains through IPR in the global context are focused on by several scholars (Krishna Kumar 1994; Gadgil et al. 1993; and also different issues of *Down to Earth*, *Honey Bee* and *Wasteland News*). The success or effectiveness of these initiatives, currently in their early phase, calls for basic changes in thinking on public interventions designed to manage natural resources with concern for local communities and local environment.

Literature on biodiversity and the community's management of it indicated a part of the process of evolution of thinking and research, beginning from enquiry into the impact of and adjustment to periodic droughts and ending with the assessment of long-term sustainability of production systems and their resource base in the fragile zones. The major inferences from this evolutionary process of understanding (as revealed by the successive literature indicated above), are as follows:

(a) The current survival and long-term sustainability of production systems and their natural resource bases in fragile resource zones are very closely linked.
(b) Enhancing and harnessing of complementarity between CPRs and PPRs can prove an effective step towards reconciling short- and long-term concerns of natural resource use.
(c) An approach to satisfy these requirements should be founded on the combination of: public interventions which involve community participation and are sensitive to the problems of fragile zone; a functional mix of technologies and institutional arrangements; which in turn are sufficiently responsive to ecology- and equity-related imperatives of the environmental and socio-economic circumstance of fragile resource regions.

A beginning on the above fronts calls for a paradigm shift, where the established wisdom of mainstream discourse, which disregarded the above perspectives, could be questioned. Accordingly, by building on the available evidence and insights one should (a) question the established view that poverty and population are prime movers of environmental resource degradation; (b) highlight the relevance and usability of social system–ecosystem links (past and potential ones) which can offer useful insights for designing and implementing strategies for resolving the inequity and exploitation-centred conflicts and dilemma characterizing the political economy of natural resource use; (c) initiate action on this front by mobilizing the experiences of community-based participatory approaches to protect the community's productive environment and livelihood against the mainstream thrusts, promoted and supported by global processes. The global processes are represented by international conventions and treaties on the conservation of biodiversity, global-warming-induced climate change, economic liberalization and impositions by WTO, etc., where marginal entities such as fragile areas and their inhabitants and the poor in general, would have little voice and visibility.

IV

As alluded to earlier, the contents and message of the volume are addressed to multiple and diverse groups engaged in: understanding the problems and prospects of environmental sustainability and livelihood security in the fragile resource regions and those associated with designing and implementing policy-programme interventions in these areas. If one goes by the past use of the approach, evidence and conclusions of the different essays, the users included academics, policy planners, NGOs and community groups, field agencies and donors. The essays have not only been frequently referred, but the contents of most of them have been reproduced fully, partially or in modified form as policy briefs, training material, or as part of the readings on specific subjects. Some fora where these essays made a difference are given below.

(a) Repeated and focused dissemination of findings of work on famine and drought did contribute towards sensitization of agencies concerned and visible reorientation of relief policies and linking them with the development programmes in India's drought-prone areas.

(b) The drought work (carried in collaboration with natural scientists) also contributed towards the reorientation of plant breeding and agronomic research strategies (by way of incorporating explicit concern for drought issues in agricultural R and D) by institutions such as ICRISAT and ICAR. (c) The pioneering work on common property resources in India not only induced subsequent work on the subject by different institutions and scholars but also formed the basis of policy briefs, advisory inputs and training material used by agencies, ranging from grass-roots-level NGOs (in Gujarat, Rajasthan, Andhra Pradesh and Karnataka) to various ministries of Government of India as well as donors such as the World Bank, in their work relating to rural development, anti-poverty programmes and natural resources management. (d) The work on sustainable agriculture and resources management in mountain areas using 'mountain perspective framework' as an approach alerted the policy-makers and planners towards the limited relevance of externally designed interventions for mountain areas. More importantly, mountain perspective was used as a broad context while preparing Action Plan for Development of Himalayas (by the Planning Commission of India); Agricultural Perspective Plan, Nepal (by the National Planning Commissioin); Strategy for Implementation of Agenda 21 in Tibet (China) through UNDP assistance; (e) The work on socio-ecological aspects and environment-poverty linkage, etc. has thrown up a number of cutting-edge issues, which have been ignored by the mainstream discourse on the subject. Going by the responses and feedback on them from the researchers and development planners, a new constituency for revisiting the conventional approaches is slowly buildig up.

The essays included in the volume were written and published during different years over more than two decades, but their relevance remains undiminished, as indicated by the focus of more recent literature on the subjects concerned. While the subjects covered range from drought management to social dimension of biodiversity conservation, or from sustainability of fragile zone agriculture to driving forces behind environmental resource degradation, the issues focused reveal strong linkages between the essays, which in turn once again demonstrates the need for an integrated approach to deal with the environmental and socio-economic problems of fragile areas. Viewed differently, the sequence of essays also reveals the history of gradual evolution of focus, ideas, and approaches towards the challenges and opportunities in fragile resources areas.

The approaches and methodologies underlying the reported work have been unique in several ways. First, the most cases it started with focus on micro-level empiricism rather than received theories. The methods often included tools and techniques used by economists, anthropologists, historians, agronomists, and climatologists. There has been a good mix of input from social scientists and natural scientists as well as policy-programme practitioners, including line agencies, project managers, NGOs, trainee groups, through feedback on the initial drafts of the essays and the ideas they contained. Viewed this way, the present volume is a product of the collective effort of many.

I. Towards Abundance

1

Famine Policies: How Apt?[*]

Drought and famine, recurrent phenomena in India, have evoked considerable attention in recent years,[1] partly on account of the state's extended involvement in famine relief. The government has often been blamed for either not having done enough or for having overdone famine relief. Morris David Morris (1975), for example, argues that expenditure on drought/famine relief involving resource transfers from equally needy non-drought-affected areas is unwarranted. Administrators, including political decision-makers, he maintains, fail to understand the intensity of a particular drought in the absence of objective criteria. The state makes its relief decisions guided (or misguided) by phenomena like migration, sale of assets, etc. by the drought-hit people. These are not, in Morris' view, distress signals actually, but rather are a part of people's adjustment mechanism to recurrent droughts. Farmers in drought-prone areas have adapted their economy to the weather cycle in such a way that, on balance, the impact of adverse years is mitigated by the impact of good years. Seasonal migration or sale of assets (accumulated in good years) during a drought year is part of the adjustment mechanism evolved by farmers. Hence the state need not respond to signals indicating depletion of stock of assets or fall in incomes. Instead relief should be provided when consumption falls below some

[*]The present essay is based on the study conducted during my tenure at CAZRI and the University of Jodhpur. I am grateful to ICRISAT for providing facilities for writing the essay and to J.G. Ryan, H.P. Binswanger and other colleagues for comments on an earlier draft. First published in the *Economic and Political Weekly*, Mumbai, under the title 'Famine and Famine Policies: Some Empirical Evidence', *EPW* 10(41) (1975).

[1]The *Economic and Political Weekly* has published quite a few contributions on the subject over the past few years. See, for instance, *EPW* 1970, Limaye and Rahalkar 1971, *EPW* 1971, N.K. Singh 1972, Mody 1972, *EPW* 1972 a, b, c, d, Wolf 1973, Patil 1973, *EPW* 1973a, b, Prabhakar 1974, Morris 1974, 1975, Rangaswami 1974, 1975, *EPW* 1975a, b, c.

level of calorie intake which, according to Morris, is the true signal of distress. Protection of consumption rather than incomes and assets should be the objective of relief policies. Morris suggests, in addition, diversification of relief projects to cover sectors other than agriculture in order to increase the employment and development impact.

Morris' prescription needs to be examined against the reality in drought-prone areas. This essay examines his argument with a micro-level illustration,[2] using evidence largely from western Rajasthan which experiences drought more frequently than any other meteorological division in India and about which I have a fair amount of first-hand knowledge and experience.[3] Does the administrator, including the political decision-maker, really respond to distress signals (which may not be genuine)? Are phenomena like sale of assets and migration true signals of drought distress? Can these signals be ignored with impunity? Can relief policies guided by protection of consumption rather than of income and assets, complement or encourage the farmers' own adjustment mechanism? The present essay attempts to answer these questions.

ALERTNESS TO DISTRESS SIGNALS

We need not refer to the provisions of the famine code or drought-determining meteorological models, whose inherent limitations as effective guides for drought relief programmes have already been brought out, but only look into the actual mechanism of state response to famine, i.e. from the declaration of scarcity to the provision of relief supplies.[4] The whole problem will be discussed in terms of the

[2]Analysis of weekly rainfall data for 47 years (1908–66) in this region showed the probability of more than one severe drought in a period of five years. Berar is another meteorological division comparable with western Rajasthan for frequency of droughts. For details see Malik and Govindaswamy 1962–3.

[3]The material in this essay was collected largely during field work for the following studies during 1963–6: (a) 'Economic Consequences of Famine in Rajasthan Desert', a study initiated in 1963–4 at the instance of late Prof. S.D. Derashri of the University of Jodhpur. The study was abandoned after collecting preliminary data; (b) Pilot Survey of Salawas Village by the Central Arid Zone Research Institute (CAZRI), Jodhpur, 1963–4 to 1965–6; (c) Jodha 1968. The first study could not be completed. The other two studies did not have drought or famine as their central theme, but drought being an inseparable feature of the economy of the region, data relating to any aspect of the economy could be easily used to reflect the impact of drought.

[4]See, for instance, Rangaswami 1974, 1975.

devices adopted in practice for determining drought intensity, etc., by various functionaries dealing with the drought problem, based on the material collected for the study 'Economic Consequences of Famine in Rajasthan Desert'.[5]

The drought and scarcity reporting system evolved during the princely rule (and which, unless diluted by other factors, works today) in the dry parts of Rajasthan bestowed a key role on the village *patwari* in assessing the situation.[6] He was responsible for evaluating the scarcity situation (and its intensity) in the village, and reporting this to the *tehsildar* who then communicated with the *Pargana Hakim* (district collector). Reporting about rainfall and crop conditions (even in the absence of drought) was part of the patwari's routine duty, but the frequency and details of his reports increased once crop failure was in sight. The reports, even in the absence of exact measurements, refined statistics and a degree of formalism, tended to give a fairly accurate assessment of the village economy at a particular juncture. The report, *inter alia,* touched upon crop prospects in terms of the *anna wari* system; the stage of crop failure (at weeding, flowering stage, etc.); whether all crops or only early/late sown crops failed; whether only grain harvest or fodder was lost; situation of top feed (*loong, pala,* etc.) from trees and bushes; availability of *pachhasas* (fodder stocked from last crops) in the village; expected duration (months) of water availability in village ponds or the biggest tank in

---

[5]The economic consequences of famine was first chosen as the theme for my Ph.D. thesis. One hypothesis was that the frequency in the region has increased. It was further hypothesized that the increased frequency of famines is due to either (a) changes in the weather pattern over the years; (b) reduced productive capacity of land resource base; or (c) changes in the methods employed for assessment of famine conditions. This required collection of some historical data, and comparing them with the present situation. To get a better insight into the past and present methods employed for determining famine situation, thirty-two serving revenue officials (ranging from patwari to district collector) with ten to fifteen years experience in drought-prone areas and thirteen former revenue officials who worked during the princely rule in the former Jodhpur state were interviewed. This was supplemented by field observations and data extracted from various records and reports of revenue officials in a cluster of four drought-affected tehsils in Jodhpur, Barmer and Pail districts (see Table 1.1).

[6]This feature may also work against an effective functioning of the system if the reporting individual fails to play his role objectively. If the patwari (or other official verifying his assessment) happens to put an informal price-tag on his service to the drought hit, he may inject a number of distortions into the system. For instance, see *EPW* 1975c. Nevertheless, during the princely rule there was no incentive for a functionary to be less objective.

neighbouring villages; number of households with supplies of foodgrains for twelve, six, three or fewer months; number of households intending to outmigrate (with approximate dates, destinations and routes, etc.); number of animals being taken to the next cattle fair for sale; scarcity situation in *chokala* (i.e. neighbouring villages); the likely situation after four-to-eight months on the basis of his experience of similar years in the past and how the situation compared with the last severe drought years.

On the basis of the patwari's report, supplemented by the revenue circle inspector's and naib-tehsildar's assessment, the tehsildar presented the picture for the whole tehsil to the hakim (district collector). At the district level usually the situation with reference to clusters of villages was examined. If only a few villages in a cluster were found scarcity-affected they were ignored except for consideration of revenue postponement or remission. If most villages in the cluster were found affected (depending upon the intensity of scarcity, crop conditions in the preceding year and present situation in the neighbouring clusters of villages) more active consideration in terms of types of relief, i.e. facilitating outmigration, supplying subsidized fodder and grain, employment on relief work, etc., was given. The hakim's report was then submitted to the counsellor who referred it to the Maharaja for final decision. Depending upon overall relief needs and available resources for the state as a whole, decisions about declaring scarcity, sanction of relief and relief measures were made.

Four distinctive features of the system of famine declaration described above may be noted.

(a) It relied heavily on the knowledge and perception of the person who had (by residence and work) intimate knowledge of the village situation. Throughout, it involved the judgement and discretion of revenue officials (who alone had traditionally been responsible for monitoring the performance of crops in different years) in assessing the extent and intensity of scarcity.

(b) The concern was less with the pattern and degree of rainfall *per se*, and more with its consequences in terms of economic conditions of the affected people. The impact of past years was also considered.

(c) It evaluated the scarcity situation for the whole area by building up evidence from below.

(d) The strong emphasis on the spatial and temporal aspects of scarcity (as reflected in the evaluation of the situation on the basis of clusters of villages and consideration of past crops) was fully in accord with the adjustment mechanism of farmers. When the situation was not uniformly bad in the whole area (and there were no successive bad years), farmers of scarcity-affected pockets could sustain themselves through migration and other mechanisms and hence there was little need for state relief.

Because of these features (particularly the last) administrators could distinguish between *kurare* (mild drought) and *akal* (severe drought), and provide selective and discriminatory relief accordingly.[7]

Nothing inherently prevents such a system working successfully even now. Revenue officials follow, by and large, the same procedures and can differentiate scarcity situations requiring public relief from those that do not.

Nevertheless, changes have come about in the objective circumstances during the last two decades or so, which directly or indirectly influence the evaluation mechanism. The change began with the involvement of other functionaries, including politicians in famine-related decisions, who had little knowledge of and direct concern for the drought problem. In the democratic context, the involvement and role of such decision-makers becomes more pronounced. In time, methods of famine determination changed and drought itself tended to become a non-meteorological and a non-economic phenomenon. Drought frequency, spread and intensity progressively became a function of irrelevant calculations at different levels of decision-making.[8] The revenue officials' function, as one of them put it, was reduced to furnishing statistics to fit drought decisions already taken. Consequently, intra-cluster and inter-cluster differences in crop conditions, which formed the relatively objective basis of scarcity or famine assessment in the past, disappeared, contributing to the state's inordinate response by way of famine relief.

Illustrative of this inordinate response are data presented in Table 1.1. The table reclassifies 195 scarcity-declared villages in four

[7]It is largely because of such selectivity of the scarcity determination system that of 56 years of adverse weather conditions between 1883 and 1939 only 29 were treated as severe drought years. The remaining 17 were considered as *kurare* (moderate drought or mild scarcity years), implying spatially uneven spread of drought conditions leading to reduced severity of the scarcity situation. For details, see Rai 1942. This rare book, a report of a senior revenue official to the counsellor to the Maharaja of Jodhpur, was available with one of the former officials interviewed.

[8]*EPW* 1975c; Rangaswami 1974, 1975.

Table 1.1 Number of Scarcity-affected Villages in Parts of Rajasthan (1963–4)[a]

Scarcity Assessment Criteria	Villages by Intensity of Scarcity Indicated by Type of Relief Qualified for							
	Land Revenue Suspension[b]	Land Revenue Remission	Fodder Supply	Foodgrain Supply	Migration Assistance	Free Water Supply	Relief Work Employment	Villages Covered[d]
(i) Villages officially declared scarcity affected	147	48	76	76	39	19	76	195
(ii) Villages scarcity affected according to 'cluster' criteria[c]	80	4	12	21	22	5	18	84
(iii) Difference (i)–(ii)	67	44	64	55	17	14	58	111
(iv) (iii) as % of (i)	45.6	91.6	84.2	72.4	43.6	73.7	76.3	65.9
(v) (iii) as % of (ii)	83.8	1100.0	533.3	261.9	77.3	28.0	322.2	132.1

Notes: [a]Data pertain to 195 villages in parts of four adjoining tehsils in Barmer, Jodhpur and Pali districts respectively. Data extracted from various records, reports and discussions with officials at the concerned tehsils and panchayat samitis.

[b]Includes suspension of repayment instalments of cooperative and *taccavi* loans.

[c]The cluster criteria of scarcity assessment mean the situation where virtually all villages in the cluster are affected by scarcity. Hence severity of distress in such cases is high because adjustment through migration and other devices is difficult. The official assessment of scarcity reflected by (i) and (ii) usually ignores the 'cluster as a unit and hence it is more liberal. It also covers drought-affected villages which (because of good crops in neighbouring villages) would have managed without public relief.

[d]Total of all columns may exceed this as one village may figure under more than one category.

adjoining tehsils in western Rajasthan during 1963–4 as either those exhibiting scarcity conditions but belonging to clusters where 80 to 90 per cent villages were free of scarcity, or those qualifying as scarcity-affected villages on the basis of scarcity-affected cluster criterion. By this criterion, nearly 57 per cent of the villages declared scarcity-affected would not have so qualified. Owing to fairly good crop conditions in neighbouring villages, they would probably have fended for themselves. Administrative ignorance would hardly explain such eagerness to declare a cluster scarcity affected, since most of the details are collected on the basis of revenue circles (i.e. groups of adjoining villages). This brings us to the phenomenon of multiple uses of drought, such as the possibilities for escaping the liability of foodgrain procurement, obtaining increased resource transfers from the Centre, and having an excuse for poor development performance of the state.[9]

Diversion of drought relief resources to unwarranted channels is another important 'use' of drought relief. On this issue the erring governments have often received strong strictures from central fact-finding teams, yet no quantified details of leakages are readily available. Table 1.2 illustrates a micro-level situation, relating to scarcity relief work at one location in western Rajasthan during 1964. Depending on the item, the leakage ranged from 23 to 67 per cent. Assuming a wage rate of one rupee per day, diversion of cash wage fund alone, due to over-reporting of both work days and the number of workers on the muster roll, amounts to Rs 55,099 (95 x 935 – 73 x 462).[10]

To reiterate, administrators are aware of the scarcity evaluation

[9]Ibid.

[10]During the 1969–70 drought, the Collector of Barmer district visited various relief works during a week and found that, in most cases, the number of workers in the muster-roll exceeded the population of the areas by 50 to 70 per cent. He ordered the closure of all relief works in certain tehsils and initiated supply of relief (part loan, part dole) on the basis of ration cards. This method of relief proved economical and effective. Later, he was transferred to a district in south-eastern Rajasthan, where droughts are uncommon.

One may probably go to the extent of recommending 'leakage proof nature' as the most relevant criterion for choosing relief measures. Present relief measures provide ample scope for leakage, but unfortunately the ones suggested by Morris, including adult literacy, training for rural craft, etc. to drought-hit people may prove more prone to leakages. At least the experience, particularly in terms of diversion of resource, of the night schools programme for rural illiterates during the 1960s and training in settlement programme for nomadic blacksmiths (*Gadoliya Lohar*) in the late 1950s in Rajasthan do not inspire much confidence in the proposals suggested by Morris.

Table 1.2. Details of Famine Relief Work at One Location in Rajasthan, 1964

Item 1	In the Records 2	In Practice 3	Difference (2–3) 4	(% leakage) 5
No. of days relief work operated	95	73	22	(23.2)
Average number of workers employed per day (for 73 days)	935	462	473	(50.6)
Road construction per worker per day (feet)	1.2	2.5		
Total cash wage payment per week (Rs)	4907	2569	2238	(45.7)
Total grain (wage) payment per week (qntl)	36.7	16.2	20.5	(55.9)
Expenses on water and other facilities per week (Rs)	115	35	80	(69.6)
Average earning (cash+kind) per worker per month (Rs)*	45.60	37.30	8.30	(18.2)

*Based on details from 113 labourers only.
Source: Adapted from the *Economic and Political Weekly* 10(19), 1975.

system which, though not foolproof, has enough potential to make their decisions more selective and discriminatory. But if for a variety of reasons they do not operate the system effectively and continue to overemphasize the other uses of drought, and if there are wide discrepancies between estimates of the extent of scarcity made by field-level functionaries and high-level decision-makers, no distress signals, whether existing or those suggested by Morris can help improve famine policies. Morris indirectly acknowledges these facts while noting that *'there is a cluster of factors mostly associated with the increasing democratization of political pressures within the system which tend to deprive famine policies of economic rationality'* (emphasis added); but attributing the deficiency to the administrators' supposed ignorance of the true nature of distress signals brushes these factors aside. Dilution of economic decisions with extra-economic considerations and the tendency of programme operators to misdirect the benefits of public goods seem to overarch operating mechanisms in a resource-scarce economy.

NATURE OF DISTRESS SIGNALS

Migration, sale of assets and other conventional signals are no doubt part of the adjustment mechanism, but the question is at what level of distress they become operative. Knowledge of this alone can help in calibrating their value as true indices of distress and so determining the stage at which state relief is warranted. This calls for a detailed study of the components of the adjustment mechanism and the sequence of their resort. In western Rajasthan, the sequence may be broadly grouped into five rungs of ascending distress:

(a) restructuring of current farm activities to maximize effective availability of products (including a variety of salvage operations);[11]

[11]Restructuring of current farm activities, to maximize effective availability of products during the scarcity year, starts as soon as the prospect of crop failure is foreseen. Accordingly, supplementary operations are undertaken besides the core operations relating to crop and animal enterprise. For example, while doing field cleaning or weeding as core activities for a crop, other activities like collection of cleared material as fodder or fuel (about which nobody bothers during a year of plenty) are performed. Such activities, leading to better management and effective utilization of available resources, are justified because of the reduced opportunity cost of labour and increased value of products thus salvaged during a drought year. Collection of rough fodder and bushes; better processing of stalk (through chalf cutting) before feeding it to animals; more discriminating

(b) minimization of current commitments through diminution, cancellation and full or part suspensions of allocation of resources;[12]
(c) disposal of inventories of home produced goods as well as purchased goods stocked for some planned use such as marriage, etc.;[13]
(d) sale or mortgage of assets;[14]

---

grazing and feeding practices involving more labour but ensuring better use of fodder/forage; collecting every piece of dung and its conversion into dung cakes, are other normally low-value activities, which get prominence right from the beginning of the period.

[12]Restructuring of intra-farm allocation of available resources takes place in the form of various activities de-emphasizing current consumption and protecting potentially productive enterprises. Reducing the family's food consumption while increasing the extent of stall feeding for increased number of animals (including currently dry cattle) in some cases, non-milking of wet animals to permit adequate milk for young calves, according higher priority to purchase of feed and fodder out of borrowings or sale proceeds of inventories or assets, ploughing back of virtually the whole of the returns from milk production for sustaining the animal enterprise, inclusion of items like gur and oil (which are almost completely dropped from the human diet) in the feed for needy animals, are a few of the outstanding examples of the tendency.

[13]Disposal of inventories to meet current requirements is a well-known practice for adjusting to the scarcity situation. Availability of stocks and adequate demand for the stocks between them determine how far sale of inventories could help the farmer during a drought year. Fodder and grain are usually not available for sale, except for big farmers (whose adjustment process is qualitatively different and in several cases they benefit considerably during scarcities). The items mostly sold in instalments—generally to the neighbouring town or to the local merchant, acting as agents of traders in the town, are: fuelwood (at times by destroying old farm fences and field shelterbelts); dung cakes, (stocked during the past one to three years); timber (preserved for own implements, cots, etc.); ropes and mats made out of goat and camel hair; spun wool kept for weaving blankets, etc.; babul tree-bark and some wild flowers (used for making liquor); ghee (stored since the last monsoon); pickles and boiled-and-dried vegetables processed out of seasonal fruits, stocks of provisions, clothing, etc. stocked for planned ceremonies.

[14]Mortgage or sale of an asset takes place depending upon its liquidity and productivity. Farmers prefer mortgage to sale. Reinforcing this preference is the low value of assets during drought and the borrower's hope of regaining a mortgaged asset on the one hand, and the lender's hope of gaining more by first mortgaging, and subsequently buying, the asset. Among assets, the mortgage of unproductive assets (like ornaments, utensils, as against land, implements, etc.) is preferred by farmers. For livestock sale rather than mortgage is common, because the carrying or keeping cost of a mortgaged animal may exceed interest to the lender. The farmer avoids selling female stock, a source of future income. Male animals are mostly sold in cattle fairs traditionally arranged at different times during a crop-year to facilitate adjustment to varying requirements of weather and crop situation.

Borrowing (mostly in kind), through land contracts and labour contracts,

(e) outmigration with animals, etc.[15]

Quantitative details on the pattern of use of these devices over time are presented in Tables 1.3a, 1.3b and 1.3c. The tables are based on data collected from a village in Jodhpur district during the drought year 1963–4 (and some comparable details for the following year).[16]

respectively in a variety of share-cropping and attached-labour arrangements, also takes place. Though exploitative (as it may permanently convert a small farmer into a tenant or an attached labourer), it is one way of adjustment where drought-hit small farmers share the protection of those who are not affected by drought.

[15]Outmigration of scarcity-hit farmers (if we ignore outmigration of artisans and landless labourers to neighbouring towns) takes place in four forms, always in groups: (a) farmers move out to irrigated areas (or other areas unaffected by drought) to work with their own bullocks and labour as share-croppers; (b) farmers with their bullocks go to towns to engage in transport activity; (c) youngsters move to irrigated (or non-drought-affected) areas as gang labour during the seasons of peak labour demand for a specific period (e.g. during wheat harvesting in Punjab and Haryana); (d) farmers migrate with their animals to Gujarat, Madhya Pradesh, Punjab, Haryana and south-eastern Rajasthan, where pasture and employment for labour is available. The last form of outmigration, called *gol*, is the most important and common in western Rajasthan during drought years. It is primarily for the maintenance of the family's capital stock, i.e. livestock, whereas the other forms of outmigration are primarily for supplementing current income through increased productive utilization of family resources. The latter are usually considered activities of relatively resourceful people having outside contacts, etc. Only a few male members of the family are involved in the first three categories: the fourth category usually involves outmigration of the whole family.

Gol achieves the advantage of spatial mobility (to adjust to the spatial variability of drought situation, possible in the case of livestock enterprise as against the crop enterprise). A household's decision to outmigrate depends upon several considerations, such as: the possibility of sustaining the productive capacity of the enterprise (in terms of length of current lactation period of milking cattle and certainty of next calving, etc.), through owned or borrowed (including public relief) resources; extent of scarcity situation in the neighbouring villages; fodder-water situation on the migration route and at the destination; the number of households from own or neighbouring villages likely to accompany; the possibility of arranging to send animals out with an outmigrating household; etc.

[16]The data in Tables 1.3a–c were collected from the Salawas village in Jodhpur district, where I worked for CAZRI from 1963–4 to 1965–6 (see n. 3). The study involved ten to twenty days stay in the village in a month. The details collected covered various socio-economic aspects of rural life, ranging from the *jajmani* system and indebtedness to the adoption and economics of conservation practices and other innovations in the village. Depending upon the nature of the problem, data were collected for the whole village or from specific samples of households. Details presented in Tables 1.3a and 1.3b were extracted from the data thus collected. Data in Table 1.3c were specifically collected to examine the impact of the time-lag between declaration of scarcity in September 1963 and

The tables cover only seven months, i.e. the period between the declaration of scarcity and commencement of relief work, and do not reflect the adjustment devices adopted by farmers in the later phase of the year. Yet they serve as a useful micro-level illustration of the emerging scarcity situation in 1963–4. Tables 1.3a and 1.3b directly or indirectly suggest curtailment of current consumption and other commitments effected by drought-hit farmers; the details in brackets describe the situation during the normal year following the drought year. Table 1.3c on the other hand suggests the devices the farmers adopted for maintaining their own low level of consumption and maintaining their livestock, etc. An interesting feature of the tables is the temporal spread of the adjustment devices adopted. Viewed in terms of various magnitudes, i.e. number of households, quantity/ money value involved in various adjustment devices during different months, it is clear that curtailment in current consumption (and other commitments) is resorted to in the early phase of the scarcity period, accompanied by sales of inventories. Once these devices are almost exhausted, farmers resort to mortgaging and, in a few cases, sale of assets. The final phase of the adjustment process consists of outmigra- tion which, according to the tables, starts much later in the scarcity period. In some cases this device is adopted simultaneously with mortgaging of assets (partly to facilitate outmigration) but in others it takes place when mortgaging as an adjustment device has already exhausted its potential. This sequence of the adjustment mechanism has a threefold implication, as follows.

First, since the intensity of distress (following the crop failure and in the absence of relief) is a function of time, the order of adjustment devices resorted to serves as an indicator of the degrees of distress. The adjustment devices usually chosen during the later stage would thus

---

commencement of relief work in late April 1964. Hence details beyond April were not collected on a number of aspects. The other details have also been accordingly presented for the same period in Tables 1.3a–c.

The foodgrain consumption data (Table 1.3a), collected partly with the help of the primary health centre in the village, were originally collected for a bigger sample of eighty households with different size of landholdings. Since some households did not reside in the village for the whole period during the drought year, details are not included in the table. Similarly, the table excludes the details collected from landless labourers as this table described only the cultivators' adjustment mechanism. The consumption details were collected by recording (in some cases by actual weighment of) the actual foodgrains consumed for one day each week. Thus the daily per capita consumption, presented in the table, is an average of four (or five) days in a month.

Table 1.3a. Varying Patterns in Farmers' Foodgrain Consumption in Drought Year 1963–4[a]

Daily Per Capita Foodgrain Consumption (gms)	Percentage of Households During						
	Oct 1963	Nov 1963	Dec 1963	Jan 1964	Feb 1964	Mar 1964	Apr 1964
300–450	7.7	21.3	34.6	48.1	57.7	60.5	69.2
	(-)	(-)	(1.9)	(5.7)	(3.8)	(5.7)	(7.6)
451–600E	21.2	25.0	38.5	34.6	35.5	32.7	25.0
	(67.3)	(73.0)	(69.2)	(74.9)	(76.8)	(78.7)	(78.7)
601–750	71.1	53.8	26.9	17.3	5.8	5.8	5.8
	(32.7)	(27.0)	(28.9)	(19.4)	(20.4)	(15.4)	(13.9)

[a]Details relate to a small sample of 52 households; Table 1.3c relates to a separate bigger sample of 144 households. Table 1.3b covers the total observations in the village as a whole. For methodological details, see n. 16. The figures in parentheses indicate the corresponding details for the year following the drought year.

Table 1.3b. Curtailment of Current Activity Commitments by Farmers in Drought Year 1963–4[a]

Items	Oct 1963	Nov 1963	Dec 1963	Jan 1964	Feb 1964	Mar 1964	Apr 1964	Total
Percentage of milk produced sold[b]	32.5 (35.7)	33.2 (39.4)	56.1 (39.4)	87.3 (32.8)	86.5 (33.1)	86.9 (35.5)	88.4 (28.4)	-
Average daily (cash+kind) sales of provisions from village shops (Rs)[c]	117 (185)	125 (490)	88 (212)	67 (220)	23 (215)	61 (368)	26 (220)	72 (245)
No. of children withdrawn from school[d]	9 (11)	7 (5)	54 (-)	42 (4)	2 (-)	2 (-)	1 (-)	117 (16)
Percentage of households which paid annual dues to craftsmen[e]	4.5 (41.0)	2.6 (38.5)	- (10.2)	(4.1)	-	-	-	7.1 (83.8)
Percentage of households who paid (part/full) cooperative dues[f]	- (10.5)	- (33.4)	1.5 (19.5)	2.4 (-)	- (-)	- (-)	- (3.2)	3.9 (66.6)

[a]The data relate to the total observations in the village as a whole. The figures in parentheses indicate corresponding details for the year following the drought year.
[b]Data relate to the households which regularly sold milk to be transported to Jodhpur town. Their number declined from 16 per cent in October 1963 to 76 in April 1964 as more and more milking animals became dry.
[c]Data based on the records of the three shops in the village. These include sale of cloth also.
[d]Data relate to the middle school in the village and a neighbouring high school attended by the children of the village. The figures exclude children who usually dropped out in August–September to help parents in cultivation and never returned to school due to drought.
[e]Total number of cultivating households getting services from carpenter, blacksmith and potter under the *jajmani* system (where payment in kind for the whole year's services is made at the time of harvest) was 195.
[f]Total members having dues were 126. Their total dues were more than Rs 37,000.

Table 1.3c. Indicators of Farmers' Distress in Drought Year 1963–4[a]

Items	Oct 1963	Nov 1963	Dec 1963	Jan 1964	Feb 1964	Mar 1964	Apr 1964	Total
Percentage households which sold inventories[b]	22.9	40.9	55.0	50.1	52.2	35.3	36.7	100.00
Percentage share of inventories sold[c]	13.7	16.2	19.5	20.2	13.6	8.1	9.7	100.00
Percentage households which sold assets[d]	-	1.4	-	-	2.1	0.7	4.2	8.4
Average landholding of asset-selling households (acres)	-	43.4	-	-	11.9	14.8	17.5	15.2
Percentage household mortgaged assets[e]	0.7	-	2.8	5.6	14.5	26.4	22.2	68.8
Average landholding of asset-mortgaging households (acres)	15.6	-	21.2	18.5	23.7	25.7	31.3	20.2
Average per household borrowing (Rs)[f]	150	-	310	350	280	370	315	346
Percentage households outmigrated[g]	-	-	-	-	1.4	4.2	21.5	27.5
Average landholding of outmigrating households	-	-	-	-	29.6	21.8	24.5	4.3

[a]The data relate to a sample of 144 farm households. Corresponding details for the year following the drought year were not collected.

[b]The percentage total may exceed 100 as a number of farmers sold inventories more than once.

[c]Average value of inventories sold per month was Rs 143 per household.

[d]Twelve households sold assets worth an average Rs 348 per household. The assets sold were mostly animals, particularly bullocks and goats.

[e]Most of the assets mortgaged by 99 households included land and bullock-cart and in a few cases animals. The percentage total would exceed 100 as five households figured more than once.

[f]This includes borrowing only against the mortgages indicated above. The borrowing against future labour contracts (i.e. attached labour) in 15 cases is not included. There were three cases where borrowing was free of any mortgage or labour contract.

[g]This indicates the households which outmigrated with their animals. This excludes three cases where some male members of the households with their bullocks outmigrated in October to work as share-croppers in the irrigated areas of Sirohi district and two cases where male members with their camels outmigrated to Gujarat for engaging in transportation jobs.

indicate greater distress. Sale or mortgage of assets or recourse to outmigration commences at a late stage when other devices have by and large been exhausted. These should therefore be regarded as true indicators of distress.

Secondly, the earlier a commitment (a necessity relating to consumption, production, etc.) is sacrificed (or reduced) the less important is it in the farmer's view while adjusting to drought conditions. For example, the fact that current consumption is curtailed first indicates that protection of a given consumption level is not the final goal of the adjustment mechanisms. On the other hand, postponement of the sale or mortgage of assets to a very late stage, preference for mortgaging rather than selling assets, migrating with the animals rather than selling them to maintain current consumption, etc., indicate that the protection of future streams of income is the main goal of the adjustment mechanism. If so, a famine policy directed solely to the protection of consumption is not in tune with the adjustment mechanism of farmers or their needs.

Further, to the extent that scarcity-affected farmers are more distressed when faced with the prospect of parting with the sources of their future income rather than the prospect of fall in current consumption, the latter ceases to be a true signal of distress.[17] Reliance on this signal may probably mean initiation of relief much earlier than is warranted by the degree of distress.[18]

[17]Instances have been noted where, even when the farmers had enough foodgrain to face the drought year, they had to outmigrate because fodder for their animals was not available. Of the outmigrating households (Table 1.3c) seven had own foodgrains to meet their reduced consumption needs, yet they left the village for maintaining their animals. Further, at least fifteen households (not covered by the sample) left the village after the commencement of relief work (construction of a road), because it had little to offer to their animals.

[18]This applies to the cultivating families. The situation is different for landless labourers (completely ignored by Morris). The labourers' adjustment mechanism mainly involves switching from 'free' to 'attached' labour status through grain loan contracts, besides a reduction in current consumption, splitting of the family to seek work, and outmigration (much earlier than farmers) to seek employment in the neighbouring towns. In the Salawas village covered by Tables 1.3a–c, there were only fourteen landless labour households, hence they did not get sufficient representation in the samples. The small proportion of such households in drought-prone areas is in keeping with the circumstances of the region, where prospects for landless labour are dim.

It is this landless section (along with small farmers) with few cushions to fall back upon, which needs early drought relief. Data collected in the second week of May 1964 from 527 labourers from different villages engaged on relief work

STATE-AIDED PAUPERIZATION

A relief policy directed at the protection of consumption and made operative only after the devices like sale or mortgage of assets and outmigration have exhausted themselves (assuming that it is only at this stage that current consumption reaches its lowest which, according to Morris, is the most relevant criterion for state relief), may in the long run prove self-defeating and contribute to the process of pauperization initiated and accentuated by recurrent droughts. It may prove self-defeating because it is not in keeping with the two positive features of a meaningful relief policy. The features emphasized both by policymakers and academics (including Morris himself when he stresses the need for development and expansion of self-insurance schemes) are (a) relief measures complementing the farmer's adjustment mechanism to face the scarcity situation, and (b) emphasis on the productive and developmental content of relief measures. Relief policies which ignore the primary goal of the farmer's adjustment mechanism and withhold relief until he has completely lost control over the sources of future incomes, can hardly be regarded as complementary to his own adjustment mechanism.

Regarding the complementarity of relief measures to the development strategy of drought-prone areas, suggested concrete steps often emphasize the creation of community assets like roads, conservation-oriented earthworks and minor irrigation works to increase the productive capacity of dry-land agriculture in subsequent years. Morris would add training for rural crafts, adult literacy, etc. A major and more certain source of growth is, however, completely ignored in these suggestions. This major source of growth, to be protected through timely drought relief, is the farmer's own production base, i.e. his capital assets, including livestock. This production base, though

---

in the neighbourhood of the village, covered by Tables 1.3a–c, showed that 36.6 and 42.2 per cent of them respectively belonged to landless households and households having less than five acres of land. Another 15.4 per cent had land between 5 and 10 acres. Only 11.8 per cent belonged to households having land of 10 acres and above. The small farmers (<15 acres) have an adjustment mechanism quite similar to that of medium and big farmers but their capacity to adjust is exhausted much earlier. (Morris' objections and recommendations about timings of relief response thus seem to be satisfied in such cases.) Instances of large and medium farmers not coming or sending their family members on relief work, as noted in several places, would also indirectly support this view. For instance, see some of the reports on drought mentioned in n. 1.

individually very small is much bigger and more significant (compared to the number of capital works), in the community aggregate in the drought-prone area.

When farmers fail to protect their production base during the drought year, the lagged impact of drought in terms of fall in the production or productive capacity of individual farms is reflected in various ways. Production-reducing impact of farmers converting into tenants, share-croppers or attached labourers following the loss, disposal or mortgage of land and other assets is palpable. So is the loss of draft power during the drought year, resulting in a fall in production due to non-cultivation of cultivable area,[19] delayed sowing,[20] and adoption of less intensive methods of cultivation[21] in the following year(s) despite adequate rains.

For livestock enterprises, the production and growth losses consequent on the farmers' failure to protect their animals are more clear. Loss of the productive stock built over a long period (dictated partly by biological factors in the case of home-bred stock) during a drought is a permanent loss of productive capacity. Non-conception due to under-feeding in the drought year leading to low calving rates in future years further reduces the production possibilities.[22] To this

[19]Old *khatoni* records (for ten years ending 1964–5) of the villages collected for Jodha 1968 showed that in the selected villages, the extent of fallow area during post-drought years was 6 to 20 per cent higher than the fallow during the corresponding pre-drought years. The fallowing of land, despite good rains, was large because of the loss of draft power and other capital resources during the drought.

[20]A study of 581 plots indicated that in the dry region with a very short wet period, even one week's delay in sowing may need resowing (or no crop) in 50 to 68 per cent of the plots. Such delay may cause nearly 50 per cent decline in bajra yield. For details see Jodha 1974.

[21]Discarding of summer ploughing and other soil working (to facilitate better moisture conservation), and inter-culturing (to ensure effective moisture utilization by crops) with their yield-reducing consequences, are the usual ways of adjusting draft-power shortages caused by a drought year. Devices, locally known as *dhira* (throwing seed on the land and attempting to cover it with soil by running a bush over it) or *khabuchya* (making a pit in the soil with a stick and sowing seed therein), where human labour is used as a substitute for bullocks/camels, were seen by the author in some villages of Barmer district following the 1963–4 drought. These are other examples where the intensity of cultivation is greatly reduced in the absence of draft power lost during the drought.

[22]Such lagged impact of drought on the cattle economy was clearly reflected by changes in the composition of dairy animals in the villages in north-western Rajasthan, where from the Delhi Milk Scheme collects milk.

may be added current losses due to reduced yield and shortened lactation period of milking animals.

Relief which protects farmers' command over resources and protect productive capacity of enterprises will serve both welfare and development goals. Also, if the full potential of future production possibilities (protected through drought relief) is assessed, it could even exceed effective relief resources spent. Drawing upon the successful informal and private arrangements operated by big farmers and traders (for their private gain) and the experience of scarcity-affected farmers during the scarcity period, some of the measures which may be tried out for strengthening the farmers' adjustment mechanism and making the relief expenditure more effective and self-liquidating should include facilitating outmigration (including initial cash and kind loans), protection against harassment by villages *en route*, provision of fodder depots and marketing facilities for animal products on migration route and at the destination,[23] supply of fodder and grain returnable in the coming years,[24] mobile marketing facilities for inventories and animal products sold under distress during the scarcity period, and provision of 'scarcity loan' by credit institutions on mortgage of assets.[25]

THE PROCESS OF PAUPERIZATION

Two more interrelated issues, complementary to the preceding discussion, that further question the validity of suggestions that famine policies should discount conventional distress signals—namely, sale of assets, outmigration, etc.—may be elaborated. They are (a) the process of pauperization initiated and accentuated by recurrent droughts and (b) the cumulative inefficacy of the conventional adjustment mechanism.

[23]The details collected from the outmigrants showed that they had to pay 40 to 90 per cent higher price than usual for the fodder and grain they purchased and received 50 to 70 per cent lower price for milk, ghee, wood, etc. they sold on the way.

[24]Large farmers successfully operate this system and earn substantial (as high as 80 per cent) interest in kind. For fodder this system can be operated through fodder banks and reserves. A similar idea has been incorporated in the proposed cattle colonies in Jaisalmer district, see *EPW* 1973a.

[25]Given the persistent conservative lending policies of credit institutions such a suggestion may appear unrealistic. Yet, if the successful working of the gold loan scheme of the State Bank of India in some areas is any indicator, one can certainly think of a viable scheme of scarcity loans. Loans against mortgage (unlike government *taccavi*) would induce the farmers to repay on time.

The literature on rural indebtedness in India furnishes ample evidence that the instability of rain-fed agriculture which gives rise to frequent famine conditions, is one of the principal reasons for mass poverty in many areas. The process by which even a resourceful farmer may be pauperized through successive droughts can be understood once the more enduring components of the adjustment mechanism are grasped. The adjustment mechanism during a famine cycle involves disposal of assets and inventories during the drought period and their gradual replenishment in good crop years. The process is not always smooth, the famine cycle not being a regular cycle. This precludes proper planning for replenishment of losses during droughts. The only element of certainty in the process of adjustment is that, owing to the specific conjunction of demand and supply factors, the affected farmer is a loser both while selling the assets during the drought year and buying them back in the post-drought year.

A broad idea of this tendency may be found in Table 1.4, which gives price details of actual transactions during a period covering a drought year, pre-drought year, and post-drought year, in three villages in western Rajasthan.[26] Notwithstanding the limited time-span covered, the table does suggest the adverse terms the drought-affected farmer is forced to accept while selling or buying as a part of the adjustment mechanism. High purchase prices, low sale prices, and high replacement prices (of assets), initiate the process of pauperization. The termination or accentuation of this process depends largely upon (a) the length of the drought-free period between two drought years, and (b) the productive capacity of the resource base the farmer is able to protect during the drought year. Those who fail to fully recoup their previous position during the good years in a famine cycle may find it near-impossible to regain the earlier position during the subsequent famine cycles. Further economic deterioration then becomes almost perpetual. This is reflected in the changes in the family's asset position, in the preference for low-cost low-productivity assets (e.g. goats in place of cattle) or the adoption of occupational patterns involving little overhead or even working capital (e.g. becoming casual or attached labour to generate family income). It is also reflected in consumption patterns.

[26]The data were originally collected (in connection with Jodha 1968) to understand the process of capital formation in arid agriculture. It was a chance factor (of course, quite natural in the arid zone) that one of the three years covered had very poor weather and crop conditions.

Table 1.4. Prices of Farm Assets and Products in Selected Villages[a]

Item	Prices[c] During the Years (in Rs)		
	1962–3	1963–4[b]	1964–5
Bullock (per pair)	875 (30)	431 (66)	988 (38)
Camel (per animal)	684 (15)	515 (27)	740 (21)
Cow (in milk-per animal)	422 (18)	220 (35)	612 (41)
Cow (dry) (per animal)	220 (12)	45 (44)	195 (9)
Sheep (per animal)	42(238)	17(780)	50 (465)
Goat (per animal)	35 (85)	47(205)	40 (317)
Bullock-cart (per cart)	630 (7)	380 (13)	690 (4)
Bajra (grain) (Rs/qntl)	32	51	43
Bajra stalk (Rs per 400 bundles)	1.1	29	1.1
Jowar stalk (Rs per 400 bundles)	24	57	26
Ghee (Rs/kg)	5	9	8
Milk (Rs/litre)	0.25	0.70	0.40
Dung Cakes (Rs/300)	8	5	8

[a]Data collected from three villages, one each from Jodhpur, Jaisalmer and Nagaur districts (for Jodha 1968).

[b]Year 1963–4 was a severe drought year while the remaining years were normal years in the selected areas.

[c]Prices for the first seven items are averages of the actual sale/purchase transactions involving the number of units (indicated in brackets). They include the items transacted in the cattle fairs and items purchased on credit by the outmigrating farmers during the drought year. Bullock-carts had rubber tyres. For bajra (grain), bajra stalk, jowar stalk and ghee the average price prevailing over the year is presented.

There are no time-series data to quantify this. Nevertheless, Table 1.5 summarizes the relevant details from case histories of seven farm families whose economic condition deteriorated after every drought year since 1939–40.[27] Through the loss of land following

[27]Case histories of fifteen 'poor' farm households were prepared during field work in 1963–5 (Pilot Survey of Salawas village, see n. 3). The selection criterion was rather peculiar. These were the only high-caste (*mali*) households with economic conditions and behaviour completely different from their caste fellows and similar to those of low-caste labour-class households. Some sort of an antithesis of what sociologists call the 'sanskritzation' process was visible, where with the deterioration of the material conditions, these families left their traditional lifestyle, and followed that of generally poor low-caste people. For example, unlike their caste fellows in the village they received bride-money; they abolished distinction of male and female living accommodation; they did not change earthen utensils in the house every *Aksaytritiya* (a sacred day during April May); they abolished a separate place of worshipping the deity in their houses; they

Table 1.5. Asset and Occupational Changes over Time of Poor Farm Families[a]

Severe Drought Years	Position Following the Drought Years Average per Family						
	Land (Acres)			Animals Owned		Members Working as	
	Owned	Sold/mortgaged[b]	Leased in	Cattle	Goats	Casual Labour	Attached Labour
1939-40	30.4	-	-	12	-	-	-
	(42.7)			(27)			
1942-3	24.6	5.8	4.5	16	-	-	-
1949-50	18.5	6.1	6.8	7	2	3	-
1952-3	13.6	4.9	8.3	3	5	4	1
1956-7	11.5	2.1	10.5	1	7	4	2
1963-4	4.8	6.7	9.5	2	11	6	3

[a]For details see n. 27

[b]Till (i.e. prior to land reforms) the land directly belonged to the Maharaja, as the village concerned was *khalsa* village (as against a *jagir* village. Tenants were owners of the land for all practical purposes, but they could not sell the land though they could permanently mortgage it. The lands thus mortgaged became the property of the mortgagee. In the post-land reforms period, one could sell the land but sales usually took place after the land remained under mortgage for a certain period and the farmer finally lost hope of regaining the land by repaying the debt (also see n. 28).

[c]Figures in parentheses indicate the position prior to the 1939–40 drought year.

every drought year the average owned landholding had declined—
from nearly 43 acres to 5 acres per family, over the twenty-five years.[28]
The number of cattle possessed similarly declined, from 27 to 2 per
family; and cows were gradually replaced by goats—the poor man's
cows. Dependence on leased-in land and labour income increased
too, as shown by the area leased-in and the number of persons working
as casual or attached labourers. Such case histories may not lend
themselves to ready generalization, but they surely give an insight
into the reality. Thus, pauperization takes place largely through (a) loss
of livestock resources from deaths, etc. or (b) loss of assets through
inter-class transfer of resources (i.e. land, livestock, etc., from several
drought-hit families to relatively better-off families). Thus the data
showing inter-class transfer of resource in the aftermath of droughts
could be a useful index of the consequent pauperization. Table 1.6
summarizes only land transfers between different groups in the villages
during more than ten years.[29] Ignoring land transfers due to

---

were not particular about using their own piece of smoke-filtering cloth while
using *chilam* (smoking-pipe) in a group; they ate at the *raika* (pastoral nomad)
houses; they castrated their home-bred animals; they kept goats, used goat
milk, and also ate non-vegetarian food; their womenfolk engaged in water-
fetching and flour grinding for others, their family members worked as attached
labourers for others, etc.

The acceptance of these practices by the high-caste households concerned
indicated social degradation. A closer enquiry revealed that pressure of economic
circumstances compelled the change. The year 1939–40, a severe famine year,
served as an important contextual point to activate their memory in the matter.
Some of the details given by them were verified from other elderly people of
their lineage group and people who lent money or acquired the land, etc., of the
households. The history of how these households were pushed on the path of
poverty is as follows: three because of prolonged infancies (due to the death of
all adult males) and partition of the family; one because other relatives usurped
the land; four due to indebtedness caused by famine and funeral feasts, etc. and
also due to untimely disintegration of the joint family leading to subdivision of
land, etc.; seven, due to drought (and subsequent mortgage/sale of assets and
indebtedness). No subdivision of land took place among the last category: no
disintegration of the family took place in four households since 1939–40; and in
the remaining three, disintegration of the joint family did not result in subdivision
of land and other farm assets, as those who left the family migrated to other
areas. The successive changes in the economic position of the seven households
are summarized in Table 1.5.

[28]Details from only four households who partly or fully acquired the land
lost by the seven families (and several others) show that, out of their average
land area owned (78.3 acres per household), more than 56 per cent was acquired
through purchase of lands initially mortgaged to them.

[29]These data were also collected in connection with the pilot study of Jalawas

Table 1.6. Inter-Class Land Transfers in Selected Villages, 1953–64[a]

Pre-transfer Landholding Ownership groups	Land Area Lost Through (acres)				Land Area Gained Through (acres)			
	Subdivision/ Gift	Litigation, etc.	Outright Sale	Sale after Mortgage	Subdivision/ Gift	Litigation, etc.	Outright Purchase	Purchase after Mortgage
1	2	3	4	5	6	7	8	9
No land[b]	-	-	-	-	298.2	17.9	82.3 (10.4)[c]	148.4 (112.1)
Below 10	70.2	132.3	66.0 (3.3)	249.6 (201.2)	24.1	11.3	-	13.2
10-25	77.1	36.5	78.3 (12.1)	113.2 (87.5)	54.4	41.4	32.7	13.2
Above 25	205.3	47.1	-	-	37.1	132.5	55.1 (5.0)	201.2 (176.6)
Total	352.6	215.9	181.4[d] (15.4)*	362.8 (288.7)	352.6	215.9	181.4 (15.4)*	362.8 (288.7)*

[a]Based on data from mutation records and follow-up details from the households concerned in three villages (Jodha 1968).

[b]This includes mainly (i) households which received land from subdivision of land after division (actual or on paper) of joint family, (ii) trader-cum-moneylender households who traditionally did not own land.

[c]Figures in parentheses indicate the land area transacted as a consequence of droughts.

[d]The land area marked * constitutes 40.0 and 88.3 per cent of the total of cols. 4, 5, 8 and 9 respectively.

fragmentation of holdings (due to family division) or gifts, etc., to circumvent land ceiling laws, which account for nearly half the lands transferred, it is clear that land transfers resulting from drought and scarcity account for 40 per cent of the lands transacted and registered in the mutation records of the selected villages. If land tranfers due to litigation (mainly following disputes after the land reforms laws were passed) are also ignored, droughts accounted for more than 85 per cent of land transfers. In these transactions lands were transferred from small and medium farmers to relatively larger farmers and to traditionally non-landowning moneylenders. In other words, droughts also widen inequalities in rural areas. Relief policies oblivious to conventional distress signals would accentuate this tendency. A properly formulated scheme of 'scarcity loans' could obviate this.

CUMULATIVE INEFFICACY OF THE ADJUSTMENT MECHANISM

The success of a farmer's adjustment mechanism against drought depends largely upon (a) the cushion of usable (including disposable) stocks and assets he is able to accumulate during the non-drought years, and (b) the felicity with which he is able to outmigrate with his animals during the crisis. Anything which obstructs this twin process detracts from the adjustment mechanism. There is evidence that the traditional settings, in which the adjustment mechanism worked quite well, have changed so quickly and considerably that the conventional devices for facing drought are inadequate. These changes are being mentioned not to idealize the traditional framework, but to highlight the backwash effects of recent developments *vis-à-vis* the working of the adjustment mechanism.

The foremost change affecting agriculture in Rajasthan after independence has been the initiation of new land policies through land reforms. The feature of land policies most relevant to the present discussion is the neglect of the intrinsic use-capabilities of arid lands in the execution of land distribution and promotion of different land-use patterns in the dry region. Completely disregarding the suitability of most such lands as natural pasture, more and more lands have been pushed under the plough.[30] The consequence has been a rapid

---

village (see n. 3), since acquisition of land was considered as one form of capital formation.

[30]For this reason, the area under crops in the region increased from 6.6 to

shrinking of grazing space, over-stocking and depletion of available pastures, complete disruption of the traditional system of rotational grazing,[31] reduced overall availability of fodder, and finally decline in the productivity of field crops because of their extension to sub-marginal lands.[32] The implications of these tendencies for building up stocks and assets to cushion the drought years can be easily perceived.

In some cases at least, the impact of the changes has been reinforced by what may be described as backwash effects of an increased degree of integration of rural and urban areas in the region through improved transportation facility—and hence reduced transaction cost.[33] A good part of the products—such as fodder, fuel wood, foodgrain and animal products like ghee—traditionally stocked for facing lean years now immediately find their way to the towns.[34] The adjustment mechanism has suffered adversely in the process, even though the region has gained in terms of marketing efficiency, etc. This is because products which so quickly move out of villages during good years, do not move so quickly and in required quantities back to the villages during the lean years. Outmigration, the need for which has been accentuated[35] by the shrinkage of local pastures and reduced fodder availability, has been made more difficult and costly with the commencement of the new order of village administration. When the drought-hit farmers and their animals traverse or come to stay in a particular village, the village panchayat, perpetually short of resources, may try to fleece

9.3 million hectares (nearly 41 per cent) between 1951 and 1952 alone. For details see Jodha 1970.

[31]The traditional system of rotational grazing, in keeping with the availability of pasture and stock watering points (called *tobas*), was governed by the authority of the *jagirdar*. After the abolition of the jagirdari system this practice has disintegrated. See for instance, CAZRI 1965: 46–7.

[32]Three-yearly moving average of area and yield of crops, from 1951–2 to 1963–4, showed that the area under bajra, jowar and sesamum had increased by 79, 20 and 105 per cent, while their yield declined by 18, 44 and 59 per cent respectively. Only in the case of kharif pulses—the crops best suited to sandy areas—the corresponding area and yield increased by 70 and 59 per cent, respectively. For details see Jodha 1972c.

[33]For details of such backwash effects of recent developments on pastoral nomads, see Jodha 1972a.

[34]For an illustration of such a tendency see Jodha 1974.

[35]According to the study of livestock breeders in the Anupgadh-Pugal region cited above, the extent of seasonal outmigration is constantly increasing. The percentage of households seasonally outmigrating increased from 3.5 in 1958–9 to 27.4 in 1962–3.

the farmers through taxes and penalties.[36] In certain cases, traditional migration routes have been blocked.[37] Overall, thus, the traditional adjustment mechanism against drought seems to have become less dependable, and relief policies need to incorporate new elements to restore their effectiveness.

[36]Details collected from the outmigrants during 1963–4 (Tables 1.3a–c) indicated that on an average each family had to spend more than Rs 150 to pacify the villagers/panchayats through penalties, taxes, fees, etc. on their way to (and back from) the Malwa region in MP. Those who had outmigrated nearly ten times in the past, reported that such levies have come into being only since 1956–7 or thereabouts.

[37]For instance, after the partition of the country in 1947, the traditional outmigration routes to canal areas in Sind were blocked for the drought-hit people of Jaisalmer, Barmer and Jodhpur regions. Some people, nevertheless, continued to informally migrate to the Sind area for animal grazing. After the 1965 Indo-Pak war the migration routes were sealed. Consequently in the drought years of 1968–9 and 1969–70, despite big relief arrangements a huge number of livestock perished in these areas, *EPW* 1972a.

2

Farmers' Strategy and Its Relevance for Drought Management[*]

The handling of the drought of 1987, considered among the worst ever, is claimed by the government to have been an innovative exercise. Claims have been made to unusual efficiency in resource mobilization, allocation and use, streamlining of the public machinery at all levels, and regular monitoring and impact evaluation (GOI 1989a). The claims have not gone uncontested. The publicized severity of the drought itself has been questioned (Bhalla and Bandyopadhyay 1988) as also the claims on operational handling (R. Chopra 1989; Mathur 1989). A highlighted aspect concerns frictions in centre–state fiscal relations on drought relief (A. Rangaswamy 1989).

Be that as it may. The purpose of this essay is to submit that in focusing on innovativeness, public policies have completely bypassed an important source of insights, i.e. the coping strategy of farmers who themselves might be considered practical drought managers. The essay elaborates on this strategy and its policy implications. It relates to arid and semi-arid tropical areas—which account for nearly 40 per cent of India's geographic area—where high rainfall variability and droughts are common and where I have had the opportunity to study farming systems in some detail.

The key message of this essay is that dry-land farmers (including pastoralists) do not manage drought in isolation from the overall farming systems. They have developed their own coping strategy in keeping with the environmental and resource specificities, but the operational components of this coping strategy is under severe strain

*First published as 'Drought Management: The Farmers' Strategies and Their Policy Implications', *Economic and Political Weekly* (Quarterly Review of Agriculture), 26(39) (1991). A revised version was published as Issue Paper No. 21 by Dryland Network Programme, IIED, London 1990.

due to rapid demographic, technological and institutional changes. The major policy implications are: (a) the need to revitalize farmers' strategy, relevant today as much as in the past, through technological and other means; (b) learning from farmers as drought managers, the need for public policies to integrate drought management with the overall development strategy for dry areas; (c) the possibility of achieving these two objectives if the rationales of farmers' strategy is made the explicit concern of integrated development and drought management intervention.

The following discussion, drawing on several results reported earlier, elaborates on these issues. After a brief description of the production environment of drought-prone areas, the farmers' adaptation/adjustment strategy is discussed. The ineffectiveness of this strategy in the changing circumstances is examined next, followed by a consideration of the policy implications of the discussion. This necessitates a brief digression into public intervention measures. The policy implications include reorienting public policy and strengthening farmers' strategy.

PRODUCTION ENVIRONMENT

In the dry areas, the annual normal rainfall ranges from less than 350 to 800 mm[1] and has a high spatial and temporal variation. Within the season rainfall varies significantly over short distances. Moisture availability is short for eight to ten months a year. In the absence of irrigation or moisture conservation, this limits the period of crop growth to roughly 60 to 180 days a year. Drought and scarcity always loom large, minimizing the options for inter-year and intra-regional variability of production possibilities, and individual production and narrow specialization. Combined with low soil fertility, paucity of moisture seriously limits nature's regenerative capacities, as reflected through low biomass productivity. Depending on the rainfall, ground water differences and soils, inter-regional differences may be seen in these characteristics, which have not only shaped farming systems but have generally influenced public policies and programmes in these areas, with a fundamental difference. Farmers, through different coping strategies discussed below, try to adapt both to good and bad rainfall situations whereas policy-makers and administrators

[1]The basis for range or values of different variables indicated here is the actual ranges of values within which the drought-prone areas (districts) studied fell.

often respond only to drought situations (Gadgil et al. 1988). Protection of drought-prone areas through periodic relief or through protective irrigation, etc. has been the focus of public intervention. Also, the various components of drought management strategy as well as development strategy are not sensitive to the specificities of the resource base in these areas.

FARMERS' STRATEGY

Through structural and operational features of their farming systems, dry-land farmers are able to adapt and adjust to both long-term and short-term climatic behaviour (Jodha and Mascarenhas 1985; Walker and Jodha 1986). By relating them to the specificities of resource base and production environment, one can view the coping strategy in perspective. For instance, moisture security and its management is a key strategy for sustenance. Similarly, since grain production is uncertain, the traditional sustenance efforts are linked to overall biomass availability and stability. This in turn helps in diversification through crop- and livestock-based mixed farming. Diversification, through activities with non-covariate flows and flexibility of decisions and operations, being an age-old defence against risk and uncertainty, is practised by dry-region farmers in the field of production, consumption and circulation (exchange). Finally, since the spatial and temporal variabilities of rainfall and their consequences hit individual households more than the communities and since inter-household differences in endowments and capacities can act as shock absorbers at a group level, collective sustenance forms an important defence against weather-induced uncertainties and scarcities. Their relative importance does vary according to climatic conditions. Gadgil et al. (1988) illustrate this by comparing arid districts of Rajasthan with semi-arid districts of Maharashtra and Andhra Pradesh. The importance of this strategy increases with increase in aridity and instability of rainfall. The variety of measures through which this strategy is operationalized are summarized in Chart 2.1. Quantitative evidence on the extent of these measures and their intra-regional differences is presented elsewhere (Jodha 1975, 1978, 1981; Gadgil et al. 1988).

Since most of the practices listed in Chart 2.1 are part of traditional farming systems and their support mechanisms, some relevant terms from ethno-sciences have been employed to categorize them. The categories are: (a) folk agronomy, to cover cropping and agronomic practices; (b) ethno-engineering, to cover traditional mechanical measures, including those for moisture conservation; (c) indigenous agro-forestry, involving complementary uses of annuals and perennials; (d) occupational diversity, including a range of activities and practices, often having non-covariate flows of output/income and input requirements; (e) self-provisioning systems, implying greater dependence on own inputs and outputs for production and consumption; and (f) collective sustenance, covering traditional forms of mutual self-help, dependence on common property resources (CPRs), etc.

TRADITIONAL STRATEGIES UNDER STRAIN

The measures listed in Chart 2.1 reflect farmers' experiences accumulated over generations. While some of them are still effective, others are under strain because the circumstances in which they were developed have, of late, considerably changed. The important factors which directly or indirectly influence farmers' strategies relate to: (a) technological changes, (b) public interventions (policies and programmes), (c) rapid population growth, and (d) increased role of market forces. These factors, except population growth, have made some positive contribution to the economies of dry lands, but they also have a number of side-effects which make the traditional adaptation/adjustment mechanisms unfeasible or ineffective. Some of the important changes with implications for farmers' strategies are summarized in Chart 2.2. Some general inferences from these changes may be noted.

APPROACHES WITHOUT PERSPECTIVE: GAINS AND LOSSES

During the last two to three decades, the drought-prone areas have to some extent benefited from development efforts and public support programmes in terms of infrastructure, high-productivity agricultural technologies, linkages with wider market situations and resource transfers for relief and development. This has, to some extent, reduced their vulnerability to severe scarcities. At the same time, most of the initiatives leading to these gains are fairly devoid of regional perspective. Consequently, in agricultural technology, high grain yield was focused upon, but flexibility of operations and input use, quantity of biomass, and potential for diversification and flexibility were ignored (Jodha 1986a, 1986b). Similarly, generalized institutional programmes like land reform, community development projects, panchayat system, etc. were extended to these areas, without assessing their potential impact

Chart 2.1. Farmers' Adaptation/Adjustment Strategies Against Drought and Uncertainty in a Dry Tropical Region of India

Measure (Category)	Adaptation Strategies for:				
	Moisture Security	Biomass Stability	Collective Sustenance	Flexibility	Diversification
Folk Agronomy					
Cultivars with:					
Varying maturity and combinability				✓	
Long duration; high stalk-to-grain ratio; yield stability		✓			✓
Cropping					
Mixed cropping; role of minor crops					
Spatial, temporal variations in planting	✓			✓	
Crop-fallow rotations	✓			✓	✓
Input use variations		✓		✓	
Ethno-engineering					
Tillage practices, weeding	✓			✓	
Moisture conservation/harvesting	✓			✓	
Irrigation structures	✓		✓		✓
Indigenous agroforestry					
Farm forestry, shelter belts	✓			✓	
Crop-bush fallow rotation	✓	✓	✓	✓	
Annual-perennial linkages	✓	✓	✓		
Occupational diversity					
Crop-livestock mixed farming	✓				
Premium on stable earning/remittance					
Acceptance of low-pay-off options	✓				
Diversity of asset structure	✓				

Self provisioning system
- High dependence on own resources ✓
- On-farm storage, recycling ✓
- Flexible consumption/resource use ✓ ✓
- Asset depletion-replenishment cycle ✓ ✓ ✓

Collective sharing systems
- Forms of mutual self-help ✓
- Common Property resources (CPR) ✓
- Migration, spatial linkage

Note: For further details and some quantitative evidence, see Jodha 1975, 1978, 1980 1986a, 1986b; Jodha and Mascrenhas 1985, Gadgil et al. 1988, Jodha and Singh 1989, Walker and Ryan 1990.

Chart 2.2. Factors and Processes Affecting Traditional Adaptation/Adjustment Strategies against Drought and Uncertainty in Dry Tropical Regions

Strategies	Factors/Processes/Influencing Strategies:			
Moisture security	In pockets improved access to irrigation, water harvesting; (-)a Boring/blasting/lifting technologies and mining of ground water	Infrastructure and support for irrigation development; (-) misallocation of scarce water due to water price policies, no use regulation	Crop intensification to meet rising demands	(-) Concentration on high water using crops, backlash on dry crops/ coarse grains
Biomass stability	(-) Reduced biomass due to concentration on grain crops and grain yields; neglect of resources centred technologies	(-) General neglect of biomass in R and D; deline of CPRs; pasture, forest, dev. dominated by 'technique' without institutional focus	(-) Decline of land extensive biomass oriented practices, e.g. land fallowing; CPRs privatized	(Rise of fodder/fuel marketing; (-) drain on rural resources, reduced local storage, recycling, availability
Collective sustenance	(-) Promotion of individual oriented (crop, livestock) technologies; missing institutional components in resource centred (watershed, rangeland) technologies	(-) Public relief/support systems replacing mutual self-help; formal legal, administrative norms replacing social sanctions; decline of common property resources	(-) Increased economic differentiation, factionalism, indifference to group action, collective concerns	(-) Market orientation and growth of individualism, erosion of group initiatives neglect of low pay-off but dependable options

Diversification	Gains of HYVs, etc. (-) backlash on minor crops, mixed cropping, land extensive cultivars, practices; neglect of non-crop, resource centred activities	Relief, employment schemes, special pro-grammes (DPAP), (-) Dependency on public relief, marginalisation of traditional diverse occupations	(-) Decline of land extensive activities, increased land fragment-ation, negative attitude to some traditional occupations	Integration with wider market economy; (-) Operational rigidities, new sources of risk, unfavourable terms of trade
Flexibility	(-) Reduced range of crops, technological rigidity of options, practices	(-) Dependence on public programmes and their logistic norms/rigidities	(-) Land constraint reducing flexibility of options	(-) Decline of self-provisioning and control over own decisions; dependence on rigidities of market system

(-) indicates negative side-effects/impacts.

on sub-marginal lands, common property resources, and various forms of mutual self-help, as bases of collective defence against droughts and uncertainty (Jodha 1986a, 1989a). Public relief strategies to help the drought-affected people were designed and pushed to such a level that they have more or less displaced the people's own adjustment mechanism and generated strong dependence on public relief (Jodha 1975, 1978; R. Chopra 1989). Irrigation facilities were developed in a few pockets but were allowed to be used on crops with high water requirements. In the process, dry crops suffered a setback (Jodha 1979; Jodha and Singh 1982). Market integration took place, but it had serious adverse effects on strategic self-provisioning systems and the fragile resource base (Jodha 1985a). Even special initiatives like Drought Prone Area Programme (DPAP), supposed to initiate development on the basis of watersheds and their specificities, in practice not only discarded the concept, but followed development norms and procedures evolved for other better endowed areas.

All this indicates the need for an understanding and explicit consideration of the specificities of drought-prone areas in both development strategies and drought management.

EROSION OF LAND-EXTENSIVE OPTIONS

An important factor behind the sustainability and resilience of diversified farming systems in dry areas was the relatively low pressure on land. Low man-to-land ratios helped in developing several land-extensive options, be it enterprise mix or agronomic practices or methods of resource management. Because of increased pressure on land (both human and livestock), most of the land-extensive options are less feasible now (Jodha 1980; Walker and Ryan 1990; Jodha and Singh 1989). Following of land-specific crop rotations, choice of cultivars with long duration, high stalk component and salvage potential (e.g. possibility of harvesting fodder in the event of crop failure), provision of common property land, etc., are specific illustrations. Besides population growth, agricultural R and D and the government's land policies have also contributed to it. These changes have significant adverse effects on biomass availability and diversification strategies.

DECLINE OF COLLECTIVE SUSTENANCE

Under the traditional strategies against weather-induced risk and drought, the measures designed for collective sustenance played an important role. Owing to the spatial variability of rainfall even within short distances and to inter-household resource difference, all households in a given cluster of villages are not exposed uniformly to crisis during a drought year (Jodha 1975; Walker and Jodha 1986). In such circumstances, when individual capacities are inadequate to combat drought, collective sharing arrangements help. Collective initiatives, developed as adaptation measures, were sustained by informal or formal understanding, enforced through social and at times religious sanctions. However, in the changed demographic, socio-political, and economic contexts, these social sanctions are no more effective. They are either completely discarded or become pointless rituals. Increased differentiation of rural communities, introduction of formal institutions, legal and administrative framework, individualism injected by market forces, etc. are major factors responsible for this change. In consequence, all adaptation/adjustment mechanisms that derived their strength from social sanctions and the community's collective approach are less feasible (Jodha 1985a, 1986a]. Most of the measures directly oriented to collective sustenance come under this category.

Other factors which have adversely affected collective sustenance arrangements are public policies and programmes (Walker and Jodha 1985). The small-scale and need-based mobilization of resources for intra-community transfers has been replaced with social transfers at the macro-level through drought relief (Jodha 1975). In the latter case, funds and supplies are controlled and distributed by the central or state governments through the district administration. Their focus is on transfer of relief resources (even fodder and fuel) from outside, rather than on generating, conserving and mobilizing them from within. Factors other than relief also play an important role in it (Mathur 1989; Rangaswamy 1989).

Most of the development initiatives, their technologies, and funding, are also focused more on individuals rather than on groups. Even for the development of collective assets, such as watershed, rangeland or social forestry, the emphasis is on 'technique', subsidy, and project framework rather than on people's involvement (Jodha 1988a). Finally, common property resources, which helped in collective sustenance and induced group participation in resource management, have been privatized on a large scale (Jodha 1986b). This has also adversely affected migration as a device to escape localized scarcities.

PEOPLE'S INITIATIVES VS PUBLIC PROGRAMMES

As an extension of the above point, one may refer to the emerging patterns where the state has appropriated several activities which traditionally formed part of people's own strategies. Though initiated with good intentions, the state's involvement in people's affairs has acquired its own logic and momentum, expanding to a level of virtually marginalizing people's initiatives in fields as far ranging as choice of crop or input use to preferred migration route or choice of activity during drought. Not coercion, but state support for some options as against others generates these consequences. Also, the bulk of the sustenance incomes during drought periods recorded in different areas (Jodha 1978) accrued from public relief. Besides generating excessive dependence on public support,[2] this pattern contains the seeds of its own long-term unsustainability. Expansion of the relief department bureaucracy and the ever-increasing need for relief funds in different states (Rangaswamy 1989; Ganapathy 1989) are pointers to this.

RIGIDITIES OF STANDARDS AND NORMS

Flexibility in resource use in production, consumption and exchange has been an important mechanism for responding to emerging situations arising from the unstable rainfall situation in dry areas. Dryland farmers operated in a largely informal social and economic environment, had farm and family as an integrated unit, depended more on self-provisioning systems and used quite flexible methods and production techniques. They are now increasingly exposed to an environment where norms and yardsticks are standardized and fairly rigid (Jodha 1981). Flexibility thus tends to decline with increased dependence on external resources and rigidities of standards or norms associated with the new technologies and support systems used by them. The new rigidities are reflected in the specificities of technological packages, terms and conditions associated with market transactions as well as support received from public agencies. These restrict farmers' capacity to change their decisions and actions as

[2]Besides the decline of the farmers' coping mechanisms, another reason for increased dependence on public relief is people's increased expectations in terms of quantity and quality of government support during scarcity. For example, drought-affected villagers, who could travel even 50 km to get relief in the past, now expect it in their own village. Earlier supply of cheap grain satisfied them, now they demand edible oil as well.

quickly as demanded for mid-season corrections to face intra-season weather variabilities.

SHRINKING BIOMASS BASE

Perennial vegetation—grass, shrub and trees—being less critically dependent on timely rainfall, their generation growth and survival are less affected by early, mid-season or late droughts than those of arable annual crops, leading to more stable production than that of crops. Moreover, the output flow of most perennials is non-covariate with that of annual crops. These factors form the basis of biomass stability and facilitate the harnessing of annual-perennial complementarities. Biomass availability from perennials is also supplemented by the choice of crops and cultivars with high stalk content. The livestock component of the farming systems, which due to its mobility characteristics already has a greater capacity to respond to spatial variability of rainfall, is used as an agency to convert biomass availability into economic gain. Livestock also plays an important role in biomass-based nutrient recycling.

The bases of biomass production are, however, rapidly dwindling. The changed choice of crops and crop technologies with low focus on fodder and decline of fodder-fuel producing resources such as CPRs (community pasture, forest, agro-forestry systems, etc.) has contributed to this process. The new public initiatives such as agro-forestry, silvi-pastoral systems, social forestry, etc. are as yet largely confined to experimental areas or pilot projects. They are yet to make an impact at the farm level. Furthermore, fodder and fuel have become important marketable products with improved transportation linkages. This has discouraged local storage and recycling. The increasing biomass crises caused by these factors, and their implications for dry-land farmers are easily seen.

PUBLIC STRATEGY

Drought proofing through irrigation or cushioning by surplus production and scarcity management through relief supplies have been the focal points of drought management since the 1890s, with relative change in emphasis or choice of specific components from time to time. Chart 2.3 lists the milestones. Some highlights (Gadgil et al. 1988) include: (a) gradual recognition and use of non-irrigation options (e.g. dry-land technology or buffer stocks from surplus-

Chart 2.3. Major Public Interventions for Drought Management and Development of Dry Tropical Areas in India (1890s–1980s)

Drought Proofing	Period of Initiation and Focus of Measures Directed to:	Drought Relief
Protective irrigation system[a]	1880s	Formation of 'famine code' to guide relief, slow response[b]
Bombay Dry Farming Technology and Soil Conservation Works[c]	1930s–40s	Relief guided by famine code
General rural development programme: community development, land reforms, irrigation development, etc.[d]	1950s–60s	Increased frequency and coverage of relief operations[e]
New agricultural strategy with focus on HYV technology for stability and surplus generation[f] Special programmes for dry area: (a) Drought-prone area programme focused on asset creation, employment and income generation, special funding[g] (b) R and D for dry lands;	1970s	Relief supported by buffer stocks, public distribution system: Focus on employment schemes, productive components in relief work[h] Attempt to use 'resource centred' technologies in relief works, uncontrolled growth in relief spending, continued centre–state friction on relief resources
Crop technology-productivity growth by external input use[i] Soil-moisture centred technology silvi-pastoral, agro-forestry, etc.[j] Extension of dryland technologies, focus on watershed approach, attention to neglected crops, problems, through 'technology missions'[k]	1980s	Formalization of need assessment, coordination, and monitoring; using past experience; emphasis on 'productivity' and 'quality' aspects; use of NGOs

Notes: Gaps and deficiencies in the measures are indicated by serial numbers. See Jodha 1979, Gadgil et al. 1988 for details.

[a]Hydrological/physical limits; protective irrigation devolved into pockets of intensive irrigation.
[b]Responded to severe crises only.
[c]Research effort short lived; lacked high productivity component; institutional unacceptability.
[d]Extension of generalized schemes without concern for resource specificities of dry areas.
[e]Ad hocism, people's increased dependence, 'leakage' in resource use.
[f]Confined to high potential, well watered areas; by passed dry areas.
[g]Not sufficiently sensitive to specificities of dry areas; poor data (and analytical) base.
[h]Isolated from development strategy; poor accountability.
[i]Focused only selected crops and crop attributes (e.g. hybrid of sorghum, millet; focus on grain yield mainly).
[j]Location specificity, focus on 'techniques', insensitive to institutional constraints.
[k]Confined to limited areas.

producing areas) for drought proofing; (b) a gradual shift away from the largely *ad hoc* measures to more planned effort under drought relief programmes; (c) supplementing of purely welfare activities (relief supplies, etc.) by productive components (social asset creation, etc.) in relief works; and (d) improved expertise for management, coordination etc. to implement the focused programme during droughts (as demonstrated during the 1987 drought). Among the important factors contributing to this evolution are: increased awareness of potential options for both drought proofing and relief (Jodha 1989b), better understanding and quantification of drought and its implications, and lessons learned from experience in handling droughts. A gap in the evolving drought management measures, which will be commented upon, is neglect of farmers' coping strategy.

In consequence, despite a number of positive changes, drought management strategy (and also development strategy) continues to be insensitive to a number of environmental and resource specificities of drought-prone areas. A few specific aspects may be noted.

Unlike the farmers' strategy, in public policy, drought management is isolated from development strategy. Consequently, even when the expenditure on assets created and employment and income generated by drought-related activities is higher than under development projects, drought management does not find a place in planning exercises (A. Bagchi cited in Rangaswamy 1989). Also, the specific components of drought management (protective works or relief works) do not have an explicit concern for diversification and flexibility requirements. Local biomass regeneration and conservation is often replaced with dependence on biomass (fodder, etc.) from outside. Collective sustenance systems, the strengthening of which can help reduce dependence on public measures, is also completely disregarded.

Effective drought management cannot be meaningfully separated from agricultural development strategy for drought-prone areas. Just as the farmers' strategy attempts to integrate long-term adaptations to dry-land environment with short-term adjustments to specific rainfall situations, public intervention should also approach drought and non-drought situations with an integrated strategy. The farmers' approach could serve as a rudimentary model for evolving a new approach, which should be explicitly concerned with diversification, flexibility, biomass stability, etc. The specificities of drought-prone areas, and their implications should also be explicitly recognized and integrated with development options. These measures would greatly help strengthen the farmers' strategy, which in turn will complement public intervention.

REVITALIZING FARMERS' STRATEGY

Though most drought-prone areas in India are less vulnerable to drought-induced crises today than in the past. (B.M. Bhatia 1989), this is less because of the dry lands' capacity to withstand drought and more because of the country's ability to spare and mobilize surpluses to help them. Even so, is advocacy of farmers' strategy relevant? Three reasons may be adduced for this relevance. First, the environmental and resource specificities characterizing drought-prone areas, such as pattern of rainfall, low biomass productivity of nature, undependability of individual options and narrow specializations, etc. remain unaltered. Second, even at the farm level, despite better income/employment opportunities due to relief and development intervention, the capacity to withstand drought has in fact weakened. Third, unless the farmers' own coping strategy is improved, the present pattern of helping them through public relief may prove unsustainable in the long run. The rate at which the relief machinery and relief requirements are expanding are a pointer to this.

Among the different components of farmers' adaptation strategy (see Chart 2.1) there could be some which may have become redundant on account of the availability of new and better substitutes. The theoretical possibilities would include new crop varieties which produce more grain and equal or more of fodder when compared to the long-duration traditional varieties, or specific drought relief initiatives which depend on government supplies but ensure group action for watershed development leading to surplus generation, which may make traditional measures unnecessary. Identification of relevant but less feasible components is relatively easy. The adjustment measures, whose bases are completely eroded, will fall in this category. For instance, land-extensive measures like crop rotation, fallowing, provision of vast areas of CPR, etc. are less feasible at their past or even reduced scales due to increased pressure on land. The same applies to measures sustained by strong social sanctions, and informality/flexibility of decision-making. Similarly the measures whose survival is contingent on complete isolation from the market are unfeasible in today's context.

On the other hand, components involving technological elements, better infrastructure and support of formal institutional arrangements can be strengthened, more so because unlike the traditional resource of poor peasants, today's society and state are well equipped in terms of scientific knowledge and capabilities, institutional innovations, and back-up support of both skilled human and material resources. The key requirement is that any development activity using these resources, when extended to dry areas, should have an explicit concern for the rationale/goals of strategies traditionally adopted by the farmers.

While revitalizing farmers' strategy one has to be clear about the following issues:

First, the advocacy for strengthening farmers' strategy to complement the existing development initiatives is based on the premise that the latter, despite their ability to raise income and periodic surpluses, are not able to contribute sufficiently to flexibility, diversification, biomass stability, collective sustenance, etc., which are essential for survival (and growth) in the climatically unstable dry areas.

Secondly, the new options to strengthen farmers' coping strategies (flexibility, diversification, etc.) should not only be more productive to take care of increased population pressure, but should have the potential to minimize the negative side-effects of market forces, public interventions and technology on farmers' coping strategy.

Thirdly, the promotion of new options when seen in the context of public strategy for drought management and development of dry areas, would imply sensitization of public strategy to the environmental and resource specificities of these areas.

Finally, the discussion mainly indicated the approaches and direction for public intervention, which can facilitate the generation of technological and institutional options, adoption of which will strengthen farmers' coping strategy. Chart 2.4 summarizes the issues. Public interventions or measures chosen for the purpose relate to agricultural technology, resource and community-centred development programmes and drought relief operations. The choice of the three categories is guided by a twin consideration: (a) potentially viable and operational measures (unlike the ones requiring high land-to-man ratio or adherence to strong social sanctions or isolation from market) can be located in these fields; (b) it is the recent developments in these areas, which have by their side-effects, largely contributed to decline of important components of the farmers' traditional coping strategy.

Chart 2.4. Possible Approaches to Generate Options to Revitalize Farmers' Adaptation/Adjustment Strategies against Drought and Rainfall Uncertainty in Dry Regions

Area of Intervention	Aspects to be Focused to Generate Relevant Options
Technologies: Crops-centred	Crop range: multiple crop choice, minor crops, cropping systems; varieties besides hybirds
	Crops with: variable maturity, variable rate and date agronomy, high temporal and spatial adaptability, compatibility (for inter cropping, agro-forestry), drought resistance, high stalk component, suited to organic recycling Products with: high storability, recyclability, local processibility
Resource-centred	Conservation measures with multiple objectives (productivity, etc.), scale and group action neutrality, focus on legthening growing season, and possibility of mid-season corrections
Perennials	Fast growing, high ratoonability, non-complementarities with annuals: focus on bio-mass processing/storage/recycling techniques
Resources/Community-centred Development Programmes	Silvi-pastoral/social forestry, etc., related initiative: de-emphasis on 'technique', formal administration and subsidy: focus on: "user group" involvement, equity of access and gains; incentives for group action, usage regulation of CPRs, involvement of NGOs
Irrigation	Focus on low water requiring crops, arrangement for equitable access; water usage regulations
Relief operations	Strong productivity components, multiple activities, emphasis on matching contribution in any from, incentive for voluntary action, involvement of NGOs, reduced domination of formal agencies, create accountability mechanisms.

Note: For further details and some quantitative evidence, see Binswanger et al. 1980; Jodha 1979, 1980, 1988b; Gadgil et al. 1988; Walker and Jodha 1986.

In agricultural technology, the focus would be on multiplication of options, involving crop varieties, management practices, complementarities between crop and livestock enterprises, between annuals and perennials and between crop-centred and resource measures. This will help widen the scope for diversification and flexibility. The specific measures in terms of attributes of different technologies to facilitate water security, biomass stability, and collective sustenance are also listed in Chart 2.4. Some steps in this direction have already been initiated by scientists (Jodha 1989b).

As regards development programmes, resource-centred measures which focus on collective sustenance and group action are emphasized, as most of the individual-centred activities already form a part of development intervention. The potential implications of these measures in terms of biomass stability and resource management involving user groups are also clear. The reorientation of water use policies in keeping with the specificities of dry areas is also called for.

The focus of measures under relief strategies would be on a greater role for people's initiatives, increased operational flexibility and greater accountability of relief operations. The search for productive components for integration with relief works is one of the most important and difficult tasks[3] indicated by Chart 2.4.

The programme discussed above can form an integrated approach, whereby drought management and development interventions could be made more sensitive to the farmers' circumstances and environmental and resource specificities of drought-prone areas. If the rationale behind the six coping strategies of farmers is made the explicit concern of relief activities, identification and choice of productive components for relief works can become somewhat easy.

[3]During the 1987 drought the Rajasthan government tried this by the creation of private assets (irrigation, well, house construction, etc.) by engaging relief labour on them. Except wages, other costs were borne by private beneficiaries.

3

Sustainable Agriculture in Fragile Resource Zones
Technological Imperatives*

 In the fragile or marginal resource areas such as mountains and rainfed arid/semi-arid tropical plains sustainability or rather unsustainability is not a matter of probability, but an already felt reality. The negative changes characterizing these areas are a consequence of current patterns of resource use that over-exploit the resources. The important resource characteristics (often ignored while using the resources) and their sustainability implications are also discussed in this essay. After commenting on the indigenous measures and conventional development interventions to handle the constraints of fragile-resource agriculture, the essay examines the need for complementing the two. Finally, the major areas requiring research focus and the technology for enhancing the sustainability of fragile-resource agriculture are discussed.

The 'sustainability' debate has created a great deal of concern in recent years. Besides the more publicized works such as *Limits to Growth* by the Club of Rome in the 1970s and *Our Common Future* by the Brundtland Commission in the 1980s, several significant contributions have been made to the subject and have been summarized by Pezzey (1989). Despite all this, sustainability continues to be a much used metaphor, with very little progress in making the concept operational (O'Riordan 1988). The problem stems from the futuristic nature of the phenomenon and associated uncertainties; required specifications of contexts which can give operational meaning to the

*First published as 'Sustainable Agriculture in Fragile Resource Zones: Technological Imperatives', *Economic and Political Weekly* (Quarterly Review of Agriculture) 26(13) (1991).

phenomenon; and the general neglect of the intra-generational aspects while focusing on the inter-generational issues as the core of the 'sustainability' debate. Nevertheless, various definitions of sustainability, which largely describe the situation rather than define the term, do highlight some broadly common elements. The important ones, as synthesized by Pezzey (1989), are summarized below.

Conceptually, the focus of sustainability is on issues of inter-generational equity. This implies equal (or greater) availability of options, in terms of human well-being or production prospects, to future generations as compared to the present one. Theoretical possibilities of such prospects, ensurable through accumulation of capital stock and technology for use by future generations, are constrained by the capabilities of the biophysical resource base. The latter cannot be stretched or manipulated indefinitely, without initiating processes of irreversible damage. This indicates the primacy of biophysical resources in sustainable development.

The operational meaning of the term sustainability, as inferred from its definitions or descriptions, provided by ecologists, environmentalists, economists and other scientists (Conway 1985; Raeburn 1984; Tisdell 1988; Chambers 1987; Ruttan 1988; Lynam and Herdt 1988; Markandya and Pearce 1988), which becomes clearer when related to specific situations, could be as follows: Sustainability is the ability of a system (e.g. the fragile-resource agriculture) to maintain a certain well-defined level of performance (output) over time, and, if required, to enhance the same, including through linkages with other systems, without damaging what Tisdell (1988) calls the essential ecological integrity of the system. Because of the time factor involved and the system's responsiveness to changing requirements, sustainability is a dynamic (as against static) phenomenon. This distinguishes sustainability from mere subsistence and makes it compatible with development.

PRIMACY OF THE BIOPHYSICAL RESOURCE BASE

Sustainability, as mentioned earlier, is a dynamic phenomenon, as reflected through the system's responsiveness to changing requirements. In the more concrete context of agriculture in the fragile zones, this 'dynamism' translates into the capacities of production factors, mainly biophysical resources, to accommodate the increasing pressure of demand without damaging their long-term potentiality. The long-term productivity and health of the natural resource base is in turn affected by the pattern and intensity of its use. Thus, devoid of finer definitional differences, in essence, sustainability/unsustainability is an outcome of match/mismatch between (a) basic characteristics of the natural resource components and (b) patterns and methods of their utilization. The latter can change (with the changing needs or perceptions of the community), but the former is normally difficult to change unless the whole resource base is transformed.

Given its inherent characteristics, the natural resource base of a system (e.g. agriculture in the fragile areas) suits only some uses. Other uses (unless the resource base itself is modified) cannot be productively maintained without either a high degree of artificial support (e.g. subsidies in chemical, biological and physical forms) or damage to the inherent capacities of the resource base itself. In either case, inappropriate use of the resource base is a definite step towards long-term unsustainability. This problem is more specific to regions with fragile and marginal land resources such as the mountains and the dry tropical areas considered here. In such habitats the unsustainability situation emerges more quickly and in a more pronounced manner. In the natural state in these areas, the range of options ensuring a proper match between resource characteristics and resource use is very narrow. Due to human ingenuity over the generations, however, the range of options has been widened. Features of traditional farming systems in these regions corroborate this (Whiteman 1988; Altieri 1987). Nevertheless, these options, having evolved in the context of low demand on fragile resources, are becoming increasingly unfeasible or ineffective in the context of the new pressures generated by population growth, market forces, and public intervention (Rieger 1981; Jodha 1986b, 1990b). The consequent measures, such as the extension of cultivation to more fragile and sub-marginal locations; the push towards monoculture induced by promotion of selected HYV crops; or the steps leading to overstocking of grazing lands and deforestation to compensate for the falling incomes, adopted to meet the situation, often fail to match well with the constraints and potential of the fragile resources (Sanwal 1989; Jodha 1986a, 1986b). A not unexpected result is the emergence of indicators of unsustainability. In such situations the re-establishment of a 'match' between resource characteristics and their use patterns is an important step in enhancing the sustainability of fragile resources and activities, including agriculture, based on them.

APPROACHING SUSTAINABILITY THROUGH UNSUSTAINABILITY

Conceptually, this reasoning implies a change in the perspective on sustainability. Accordingly, for identifying and operationalizing the components of sustainability for a given system, one needs to examine unsustainability first and then proceed backward to understand the factors and processes contributing to it. This can help in identifying practical measures to reverse the process leading to unsustainability. A practical step towards implementing the above approach is to prepare an inventory of the indicators of unsustainability in a system and then look into the 'why and how' behind them. This approach has some merits. It can help in improving the understanding of operational aspects of the issues involved in the sustainability debate. This also helps to relate more easily the involved issues to the real world situations in which the causes and consequences of unsustainability are felt. It can also help identify concrete steps to modify the current approaches towards development and resource management. Such steps may relate to macro- and micro-level policies and programmes as well as to farm-level decisions and actions. Using this framework, we shall discuss first the indicators of unsustainability characterizing the mountains and the dry tropical regions. This will be followed by a description of resource characteristics, disregard of which, at different levels, is primarily responsible for the emerging indicators of unsustainability. The sustainability implications of the resource characteristics are indicated. This will be followed by a brief discussion of the extent to which these implications have influenced production and resource management practices under the traditional farming systems, and the conventional development approaches to agriculture in the fragile zones. The sustainability-promoting features of the two are highlighted to indicate the possible scope for blending them for enhanced sustainability of agriculture. Their practical implications are presented as the potential focus areas for research and technology development (R and D) for fragile-resource agriculture. Discussion on equally important institutional and demographic factors for sustainable agriculture falls outside the scope of this essay.

FRAGILE REGIONS: THE DOMINANT SCENARIO

The dominant characteristic of most of the fragile regions in developing countries, particularly those with high population pressure,

is the widening gap between development efforts (indicated by investment and public interventions) and the corresponding achievements
in terms of measurable economic gains (especially on per capita basis)
and qualitative changes, such as the health and production potential
of the natural resource base, environmental consequences, etc.
During a brief span of forty to fifty years, several alarming trends
have emerged in different parts of the mountains and the dry tropical
regions. There are, in these regions, clearly visible, persistent negative
changes, relating to crop yields, availability of other agricultural
products, the economic well-being of the people, and the overall condition of environment and natural resources (Rieger 1981; Glantz 1987;
Blaikie and Brookfield 1987). For instance, in mountain areas at
present, in comparison to the situation fifty years ago, the extent and
severity of landslides is higher; water flow in traditional community
irrigation systems (*kools*) is lower; yields of major crops in the mountains (except in highly patronized pockets) are lower; diversity of
mountain agriculture is reduced; the inter-seasonal hunger gap (food
deficit period) is longer; time spent by villagers for collection of fodder
and fuel from neighbouring uncultivated areas or common property
lands is longer; the botanical composition of species in forests and
pastures has undergone negative changes; and finally, the extent of
poverty, unemployment and outmigration of the hill people has
increased. Ives and Messerli (1989) refer to some of the persistent
negative changes referred to above in the Himalayan context. An
inventory of such measurable, verifiable or objectively assessable
changes in the selected hill areas of Nepal, India, Pakistan and China
has been made (Jodha 1990b).

In dry tropical areas, various forms of resource degradation
including increased salinity (of both soil and ground water), deepening
of water tables, disappearance of plants from pastures and community
forests and increase of areas under shifting sand are quite visible.
Similarly, during recent decades decline in overall biomass availability,
substitution of cattle (and camels in arid areas) by sheep and goats,
and the extension of cropping to sub-marginal areas to meet production deficits have been observed. Reduced productivity and reduced
resilience of the traditional farming systems have led to increased
dependence on public relief and increased seasonal migration to other
areas. Various facets of the decline have been recorded by different
studies (Jodha 1986a, 1988b; Warren and Agnew 1988). However,
the situation in the limited areas transformed through dependable
irrigation systems is quite different.

INDICATORS OF UNSUSTAINABILITY

These negative changes, treated as indicators of unsustainability, may relate to (a) resource base (e.g. land degradation); (b) production flows (e.g. persistent decline in crop yields); and (c) resource management/usage options (e.g. increased unfeasibility of annual-perennial based inter-cropping or specific crop rotations, etc.). More importantly, for operational and analytical purposes, the indicators may be grouped under the following three categories on the basis of their actual or potential visibility. Charts 3.1a and 3.1b summarize some of them for the mountain areas and the dry tropical areas respectively.

Directly Visible Negative Changes: In mountain areas, these may include increased landslides or mudslides; drying of traditional irrigation channels (kools); increased idle periods of grinding-mills or saw mills operated through natural water flows; prolonged fall in crop yield; reduced diversity of agriculture; abandonment of traditionally productive hill terraces; and increased extent of seasonal outmigration of hill people. In dry tropical areas, such changes are reflected in various forms of resource degradation and desertification and their impacts. Some of these include accentuated soil erosion; increased salinity of soil and ground water; increasing severity of drought-induced scarcities; reduced feasibility and efficacy of traditional adaptations against weather risks; reduced overall biomass availability; and reduced carrying capacity of pastures.

Negative Changes Made Invisible: People's adjustments to negative changes often tend to hide the latter. In both mountain areas and dry tropical areas such changes may include: substitution of shallow-rooted crops for deep-rooted crops, following the erosion of top soil; substitution of cattle by small ruminants due to permanent degradation or reduced carrying capacity of grazing lands; introduction of public food distribution systems due to the increasing interseasonal hunger gaps (local food production deficits); small farmers leasing out their lands to concentrate on wage earning; and shift towards increased external inputs in cropping due to the decline of locally renewable resources.

Development Initiatives with Potentially Negative Consequences: A number of measures are adopted for meeting present or perceived future shortages of products at current or increased levels of demand. Some of the measures (changes), while enhancing productivity of agriculture in the short run, might jeopardize the ability of the system to meet the increasing demands in the long run. Chances of such happenings are positively linked with the interventionts' insensitivity to specific conditions of the fragile-resource areas (Franke and Chasin 1980; Altieri 1987; Jodha 1986b).

This can be illustrated by any farm technology that increases crucial dependence of mountain agriculture on external inputs (e.g. fertilizer) as against the locally renewable input resource, or adds to mass production of high-weight low-value products with a largely external market ignoring the inaccessibility and related problems. Similarly, any measure that disregards the fragility of mountain slopes and ignores linkages among diverse activities at different elevations in the same valley (e.g. farming-forestry linkages) and promotes monocropping may not prove sustainable. In the dry tropical areas, any intervention that disregards the totality of the farming system (covering crop, livestock, and their support mechanisms); over-emphasizes grain yield at the cost of total biomass; focuses on high-cost inputs for low-value coarse grains; and attempts to upgrade resources (e.g. by irrigation) ignoring soil characteristics and the impeded drainage situation, etc. cannot ensure sustainability of agriculture.

Under invisible negative changes and development initiatives with potentially negative consequences, there may be several changes which might bring positive results in the long run and enhance the sustainability of agriculture in the fragile areas. To separate them from negatively-oriented changes, one needs a fairly detailed analysis of the involved components. This involves examination of the implications of interventions in terms of their compatibility with the relevant characteristics and conditions of the fragile areas, namely, inaccessibility, fragility, marginality, diversity, niche, and the adaptation mechanisms of people in these habitats. These can be used for screening the relevant interventions and sustainability implications for fragile-resource agriculture. The utility of such an effort will depend on (a) identification of factors and processes contributing to the persistent negative trends and (b) identification of measures to handle such factors and processes. Population pressure, increased role of market forces, and the side-effects of public interventions in the recent decades are often identified as basic factors causing and accentuating the negative trends mentioned earlier (Banskota 1989; Jodha 1986a, b; Grainger 1982).

At the same time, two points need emphasis. First, neither market nor public intervention (and even population growth in the near

Chart 3.1a. Negative Changes as Indicators of the Unsustainability of Agriculture in Mountain Areas

Visibility of Change	Changes Related to:[a]		
	Resource Base	Production Flows	Resource Use/Management Practices/Options
Directly visible changes	Increased landslides and other forms of land degradation; abandoned terraces; per capita reduced availability and fragmentation of land; changed botanical composition of forest/pasture Reduced water-flows for irrigation, domestic uses and grinding mills	Prolonged negative trend in yields of crop, livestock, etc.; increased input need per unit of production; increased time and distance involved in food, fodder, fuel gathering; reduced capacity of saw mills operated on water flow; lower per capita availability of agri. products; etc.	Reduced extent of: fallowing, crop rotation, inter-cropping, diversified resource management practices; extension of plough to submarginal lands; replacement of social sanctions for resource use by legal measures; unbalanced and high intensity of input use, etc.
Changes concealed by responses to changes	Substitution of: cattle by sheep/goat; deeprooted crops by shallow-rooted ones; shift to non-local inputs Substitution of water flow by fossil fuel for grinding mills; manure by chemical fertilizers[b]	Increased seasonal migration; introduction of externally supported public distribution systems (food, inputs)[b] intensive cash cropping on limited areas[b]	Shifts in cropping pattern and composition of livestock; reduced diversity, increased specialization in monocropping; promotion of polices/programmes with successful record outside, without evaluation[b]
Development initiatives, etc.— potentially negative changes[c]	New systems without linkages to other diversified activities; generating excessive dependence on outside resource (fertilizer/pesticide-based technologies) ignoring traditional adaptation experiences (new irrigation structure)	Agricultural measures directed to short term quick results; primarily product-(as against resource) centred approaches to agricultural development, etc.	Indifference of programme and policies to mountain specificities, focus on short-term gains, high centralization, excessive, crucial dependence on external advice ignoring wisdom

[a]Most of the changes are inter-related and they could fit into more than one block.

[b]Since a number of changes could be for reasons other than unsustainability, a fuller understanding of the underlying circumstances of a change will be necessary.

[c]Changes under this category differ from the ones under the above two categories, in the sense that they are yet to take place, and their potential emergence could be understood by examining the involved resources use practices in relation to specific mountain characteristics. They manifest the 'process' as against 'product' aspect of unsustainability.

Chart 3.1b. Negative Changes as Indicators of the Unsustainability of Agriculture in Dry Tropical Areas

Visibility of Change	Resource Base	Changes Related to:[a] Production Flows	Resource Use/Management Practices/Options
Directly visible changes	Various forms of resource degradation: emergence of salinity, coverage of fertile soil by shifting sands, vanishing top soils due to water/wind erosion; deepening of water tables, ground water salinization; emerging plantlessness, reduced perennials, increased inferior annuals and thorny bushes; reduced per capita availability of productive resources	Reduced total and per capita biomass availability; reduced average productivity of different crops, increased cropping on sub-marginal lands, reduced product recycling; higher dependence on inferior options, (e.g. harvesting/lopping premature trees), rising severity of successive drought-impacts; increased dependence on public relief, increased migration	Changes in land-use pattern: cropping on sub-marginal lands; decline of common properly resources; reduced diversity of agriculture (e.g. number of crops/ enterprise and their inter-linkages); reduced feasibility and effectiveness of traditional adaptation strategies (e.g. rotations, inter-cropping, biomass strategies)
Changes concealed by responses to (negative) changes	Substitution of cattle, camels, by small ruminants; increased emphasis on mechanization of cultivation and water lifting; reduced idling of land; large-scale 'reclamation' (!) of wastelands; shift from local to external inputs (e.g. from manure to chemical fertilizers, wooden tyres to rubber tyres for bullock carts)	Higher coverage by public distribution system (food, inputs) and other anti-property pro-grammes;[b] reduced resilience and greater dependence on external market sources; changes in landuse pattern favouring grain production	Discarding of minor crops, shift towards monocropping with standardized inputs/practices; increased land use intensity; shift from two-oxen to one-ox plough; replacement of tractorization[b]; replacement of self-help system by public support systems

Development initiative, etc.— potentially negative changes[c]	R&D focus on: crop rather than resource; technique rather than user-perspective e.g. method/species/ inputs, group action for watershed/ range development); resource upgrading ignoring its limitations (e.g. irrigation in impeded drainage areas); including high use intensity of erodable soils, and other resource extractive measures (e.g. tractoriz-ation)	Highly subsidized, narrowly focused production programmes: focus on crops ignoring other land based activities, grain yield ignoring biomass; monocropping ignoring diversification; relief operations focused on people and livestock ignoring resource base, thus promoting high pressure on poor resource base	Sectoral focus of R&D and other support systems ignoring flexibility and diversification needs; privatization of common property resources; extension of generalized external approaches to specific areas: disregard of folk knowledge in formal interventions; replacing local informal arrangements by rigid legal/administrative measures

[a]Most of the changes are interrelated and they could fit into more than one block.

[b]Since a number of changes could be for reasons other than unsustainability, a fuller understanding of the underlying circum-stances of a change will be necessary.

[c]Changes under this category differ from the ones under the above two categories, in the sense that they are yet to take place, and their potential emergence could be understood by examining the involved resource use practices in relation to specific area-resource characteristics. They manifest the 'process' rather than 'product' aspect of unsustainability.

future) can be wished away. Secondly, it is not so much the presence of these factors but rather their interaction pattern with the fragile resources and environments that matters.

The sustainability implications of the characteristics of fragile areas can be understood in terms of the degree of convergence between: (a) objective circumstances (e.g. operationally relevant constraints and potentialities) created by them and (b) conditions associated with the process of sustainable development (e.g. ability of a system for sustained performance without damaging its essential ecological integrity). To elaborate on this, we need to refer back to the operational meaning of sustainability mentioned earlier. Accordingly, sustainability (i.e. sustained or increased level of production performance) is conditioned by the capacities of the biophysical resource base to withstand high use intensity; to absorb high quantities of complementary inputs; to tolerate periodical shocks/disturbances without facing permanent damage; to ensure gains associated with the scale of operation and infrastructural logistics; and to gain from linkages/ exchanges with other (wider) systems.

Juxtaposition of the above requirements (or preconditions of sustainability) and the already discussed characteristics of fragile resource areas can greatly clarify the sustainability problems of fragile-resource agriculture. This is attempted through Chart 3.2 which presents a broad view of the complex of constraints and potentialities created by the natural resource base of fragile areas. It can also serve as a framework within which the search for sustainable agriculture can be directed. The major areas that need attention may be presented in the form of some focused questions.

(a) In view of the fragility, marginality, and to some extent inaccessibility problems, how can the use intensity of land and its (physical and economic) input absorption capacity be enhanced without negative side-effects, i.e. resource degradation?

(b) What are the options for developing a complex of diversified activities with clear focus on: (i) high productivity despite low land-use intensity and low input regimes (especially external inputs); (ii) fuller use of resource diversities and niches (i.e. the options with comparative advantages), without over-exploitation and degradation of resources?

(c) How to strengthen the resilience of farming systems to cope with the periodic shocks/stresses and the rapid growth of pressure on fragile-resources?

Chart 3.2. Sustainability Implications of the Specific Conditions of Fragile Areas[a]

Resource Specificities[a] (and objective circumstances)	Sustainability Implications—Possibilities through						
	a) Inherent Production Potential and Modification				b) Abilities to Link with Wider Systems		
	Resource Use Intensity	Input Absorption Capacity	Infra-structural Logistics	Gains of Scale	Resilience to Shock	Surplus Generation & Exchange	Replicability of External Experience
Inaccessibility: Remoteness, distance, closeness, restricted external linkages, etc.	(−)[a]	(−)	(−)	(−)	(−)	(−)	(−)
Fragility: Vulnerability to irreversible damage, low carrying capacity, limited production options, high overhead cost of use, etc.	(−)	(−)	(−)	(−)	(−)		
Marginality: Cut off from mainstream, limited production options, high dependency, etc.	(−)	(−)	(−)	(−)	(−)	(−)	
Diversity: Complex of constraints and opportunities, interdependence of production bases and products/activities, etc.	(+)[a]	(+)	(+)	(−)	(+)	(−)	
Niche: Small and numerous specific activities with comparative advantage; use of some beyond local capabilities, etc.	(+)	(+)[a]	(+)	(−)	(+)	(+)	(−)
Adaptation Mechanisms: Folk agronomy, ethno-engineering, collective security, diversification, self-provisioning, etc.	(+)	(+)	(+)	(−)	(+)	(+)	(−)

(−) indicates extremely limited possibilities; (+) indicates greater scope for sustainability through production performance and linkages with wider systems (e.g. upland–lowland interactions).

[a] The resource specificities or biophysical conditions indicated here are more pronounced in mountain areas than in dry tropical plains.

(d) What should be the potential forms and patterns of linkages of fragile resource agriculture with other systems (i.e. agriculture and the general economy of other zones), in order to facilitate the accomplishment of potential options under questions (a)–(c)?

PAST STRATEGIES

These issues, even though not formulated as such, have in the past been addressed in various ways. Rural communities have evolved and inherited their adaptation strategies to handle these problems. In recent decades, through development interventions, the same problems have been focused more formally. Chart 3.3 summarizes the relevant components of the two which are directed to resource management and productivity growth in agriculture. The chart reveals both the strengths and weaknesses of the two approaches. A synthesis of the strengths of the two may help in identifying the directions and possible first-order options to enhance the sustainability of fragile-resource agriculture.

By way of a comment on Chart 3.3, the following may be stated. Traditional measures and practices have been evolved by people through informal experimentation over the generations (Chambers et al. 1989b; Altieri 1987). Hence, they are better adapted to limitations and potentialities of the fragile resources. Broadly speaking, they are location-specific and small in scale; diversified and interlinked in their structure and operations; often land-extensive and locally renewable resource–centred; mainly supported by folk knowledge and informal social sanctions; and generally have lower input use and lower productivity. For the above reasons, they are conducive to sustainable resource use under low pressure of demand in relatively isolated or inaccessible situations. But they are becoming increasingly unfeasible and ineffective in the context of rising pressure on fragile land resources.

The measures promoted through conventional public intervention in fragile zones, on the other hand, generally represent the extension of land-intensive production system characteristics of relatively better agricultural areas (Altieri 1987; Jodha 1986b). So far they are not well adapted to the fragile resources. Public intervention, on its current scale and level of standardization, is a recent phenomenon in these areas. Being in the early stage of evolution compared to traditional measures, probably it can be modified to suit the situation in fragile zones. Its major strengths are the significant input of modern science and technology; strong (public exchequer-based) resource support; and conscious decisions and efforts to relax the development constraints of the fragile-resource areas. It has significant potential for strengthening physical and market linkages among fragile zones and other regions. It should, be noted however, that past efforts based on these positive attributes have not strengthened the prospects of sustainable development in fragile areas. On the contrary, several indicators of unsustainability have emerged side by side with development efforts. The primary reason for this has been the general insensitivity of public interventionists to specific conditions of the fragile zones. To impart this sensitivity, an effective approach would be to redesign the intervention by blending the rationale of traditional measures with formal technological intervention (Chambers et al. 1989b).

Viewed differently, the whole issue of sustainability of agriculture in fragile areas can be reduced to a problem of increased use intensity of land resources (for higher productivity) without permanently damaging them. The indigenous systems, though oriented to resource use with conservation, do not possess high-productivity technological components to ensure high use intensity and resource conservation simultaneously. The new science- and technology-based interventions have the capacity to raise use intensity and productivity of land but they are generally indifferent to conservation considerations. The above facts form the basic ground for blending the positive features of the two (Dregne 1983; Jodha 1986b).

To facilitate such blending, some areas or issues may be suggested, that can be focused by research and technology development (R and D) efforts. Chart 3.4 summarizes them, along with their potential impact areas, to enhance the sustainability of agriculture. Concentration on these subject-areas would require substantial reorientation of R and D strategies in the fragile zones. This would imply making research and technology measures more resource-centred, system-oriented, and conducive to harnessing the niche and diversities of resource base. The need for institutional and other logistic support to complement the technologies hardly needs mention. The links between specific technological measures and conditions associated with sustainability, indicated in Chart 3.4 are briefly commented upon here.

Resource-centred Research and Technologies: For the enhancement of input absorption capacity and use intensity of fragile resources, both

Chart 3.3.   Measures Against Constraints to Sustainable Agriculture in
Fragile Resource Zones (Indigenous Systems/Development
Interventions)

Traditional Farming Systems	Conventional Development Interventions

A. Enhancement of Use Intensity/Input Absorption Capacity of Land

Measures

Resource amendments by ethno-engineering measures: terracing/trenching/ridging, moisture conservation/drainage management/shelterbelts/ agro-forestry, etc.

Selective resource upgrading through irrigation/other infrastructure, biophysical changes (e.g. new introduction; R&D activity/pilot projects for range lands, watersheds, etc.).

Attributes Conducive to Sustainability

Local resource centred, community oriented and supported, small scale, diverse, adapted to local situation; linked to other activities.

Science and technology input, strong logistic/resource support, advantage of scale.

Limitations

Reduced feasibility with rising pressure on land and weakening of local-level collective arrangements lack new high-productivity components.

Side-effects of massive interference with fragile resources (water logging, salinity, landslides); inequities between transformed (e.g. irrigated) and leftover areas; insensitivity of R &D based initiatives to local resource diversity and user perspective.

B. Usage and Management of Low Use-Capability Lands

Measures

Folk agronomy involving activities with low land intensity and low (local and affordable) input regimes; integration of low-intensity–high-intensity land uses (based on annual-perennial plants, crop-fallow rotations, indigenous agroforestry, common property resources; seasonal migration/transhumance).

Sectorally separated production programmes; high intensity uses through new technology inputs/incentives/subsidies; focused conservartion-oriented intiatives (forest/pastures/watersheds) largely in projects mode.

Attributes Conducive to Sustainability

Diversified, interlinked activities with different levels of intensity,

New technological input, resource support and top-down legal

(Continued)

Chart 3.3 continued

Traditional Farming Systems	Conventional Development Interventions
community participation, control on local demand	and administrative arrangements.

Limitations
Reduced feasibility and effectiveness due to population growth, decline of collective arrangements, and side effects of technological and institutional interventions.

General indifference to resource limitations, user perspective; 'technique' and 'project mode' dominated measures.

C. Options to Harness Diversity and Niche

Measures
Folk agronomy—diversified cropping, focus on multiple-use species; complementarily of cropping-livestock-forestry/horticulture; emphasis on biomass in choice of land use and cropping patterns; complementarity of spatially/temporally differentiated land-based activities; stability-oriented, location-specific choices; harnessing niche for tradable surplus, food security.

Sectorally segregated programmes and their support systems (R&D, input supplies, crop marketing); focus on selected species and selected attributes (e.g. mono-culture, high grain: stalk ratio); extension of generalized develop-ment experience of other habitats with high subsidy.

Attributes Conducive to Sustainability
Diversification, linkages as dictated by resource characteristics; locally renewable resource focus.

Initiatives with strong technological and logistic components, high potential for generating new options.

Limitations
Low productivity, land extensive measures incompatible with high man–land ratio, and changed institutional environment.

Indifferent to the totality of farming system and diverse resource potentialities; high subsidization.

D. Resilience of the System and Mechanisms to Handle High Pressure of Demand

Measures
Diversification and linkages of land-

Public relief and support during

(Continued)

Chart 3.3 continued

Traditional Farming Systems	Conventional Development Interventions
based activities; flexibility in scale, operations, input use; locally renewable resource focus, recycling of inputs/products, self-provisioning; crisis period collective sharing arrangements, common property resources, social regulations for rationed use and protection of fragile resource; release of periodic/ seasonal pressure by migration, transhumance, remittance economy.	crisis/scarcities; public interventions replacing traditional self-help strategies and informal regulatory measures; highly individual (not community)-focused interventions (e.g. privatization of common property resources, crisis period cushion promoted by increased private-resource productivity by HYVs, etc.); occasional linking of relief measure with productivity measures.

Attributes Conducive to Sustainability

Range of options to match specific constraints of the habitats; emphasis on community centered and regulated activities; rationing of demand on fragile resource.	Resource transfer from better-off areas to scarcity prone areas; possibility of linking relief initiatives with resource conservation/ production programmes.

Limitations

Infeasibility and reduced efficacy of collective self-help measures and folk agronomic devices, due to changed demographic, institutional, and technological environment.	Dependency for sustenance on external resources; encouragement for perpetual growth of pressure on fragile resource; indifference to local self-help initiative.

E. Linkages with Other Systems (including Wider Market Systems)

Measures

General state of relative inaccessibility (particularly for mountains) and isolation from main stream market; limited linkages through tradable surplus from harnessing niches; crisis period external dependence through transhumance, migration, pockets of remittance economy.	Improved physical and market linkages; integration of fragile resource economy with other systems; focus on special area development programmes, transformation of limited areas and their demonstration effects.

Attributes Conducive to Sustainability

A few positive side effects of	Improved opportunities for relaxing

(Continued)

Chart 3.3 continued

Traditional Farming Systems	Conventional Development Interventions
isolation, local demand centred, socially controlled extraction of fragile resources, better links between the resource users and the resources.	internal constraints through technology, resource transfer, interactions with other systems; inducement for fuller use of niche through external demand; closer integration with mainstream.
Limitations Persistent neglect and marginal status of fragile-resource areas; slow pace of transformation of agriculture; unfavourable terms of exchange for marginal areas and products.	Unless guarded against high chances of extending irrelevant external experiences (including technologies); external demand induced heavy extraction of niche; unfavourable terms of exchange; distortion in local demand patterns and resource use.

mechanical and biological measures can be employed. Traditionally, people treated fragile resources through measures such as terracing, trenching, ridging, hedges and shelterbelts, etc., and made them usable. With better scientific understanding of the precise interactions between resource components and specific treatment variables, new and more effective options can be evolved to handle the problems of slope, drainage, marginal soils, and excess or deficit moisture (Dregne 1983; El-Swaify et al. 1985). Plants with high soil binding and soil building capacities can also form important components of new technologies. Integrated use of (i) early maturing, fast growing perennials (including trees and shrubs) and (ii) photo-period insensitive, early maturing, high-yielding annuals can be an important step for increased resource use intensity. Species with high productivity and high value, suited to fragile resources, can economically enhance their input absorption capacity (Nair 1983). With focused screening of the available vast and diverse germplasm, it should be possible to identify several species with these attributes.

A number of resource-centred technologies being developed at present implicitly focus on some of the above issues, but these initiatives, be they pasture development through reseeding or soil manipulation, etc. or the more publicized and subsidized initiatives such as integrated watershed development, are highly 'technique'

Chart 3.4. Areas of Focus for R&D for Sustainable Agriculture in Fragile Resource Zones

Areas/Issues of Focus	Potential Enhancement of Sustainability through			
	Resource Use Intensity	Input Absorption Capacity	Resilience and Productivity	Inter-systems Linkages
Resource-centred R&D				
Physical/biological measures to manage	✓	✓		
Soil binding/building plants/crops		✓	✓	
Perennials (fast growing, early maturing, high productivity, high rationality)	✓			
Biological control of yield reducers	✓	✓		
Locally renewable resource focus	✓	✓		
Location specificity	✓	✓		
System-oriented R&D				
Linkage of product and resource-centred options	✓		✓	
Diversified—interlinked activities (annual-perennial plants)	✓		✓	✓
Extensive—intensive land uses	✓			
R&D focused to harness niche and diversity				
Wider adaptability of options	✓		✓[a]	✓[a]
Focus on productivity of total system				
Flexibility in input use/agronomy				
Recyclability/storability				
Complementarity with other zones (related to input, product, value additions)			✓	✓
Input of knowledge			✓	✓
Infrastructural/institutional and resource support[b]			✓	✓

[a] Gains through advantage of scale and replicability of external experience.
[b] Options under this category may not directly relate to R&D.

dominated, and they are still conceived and implemented in 'project mode'. The institutional factors and user perspectives are almost completely neglected, and this reduces their relevance to the problems of these areas (Blaikie and Brookfield 1987).

Although due to resource heterogeneities, the location specificity of technologies cannot be avoided, emphasis on wide adaptability of technological components can facilitate wider coverage and advantage of scale to specific production activities in fragile-resource agriculture.

Systems Approach: Suggesting a greater need for a systems approach for fragile-resource agriculture amounts to stating the obvious. Yet, to avoid the gaps characterizing the conventional approach, a few issues need to be mentioned specifically.

Diversification and interlinkages of different land-based activities have been the major strengths of traditional farming systems. The linkages can be seen between the activities based on annuals and perennials; intensive and extensive resource uses; and complementary uses of common property resources and private property resources (Jodha 1986b). Diversity and implied linkages are important considerations in the choice of crops and their attributes. Moreover, in such systems, productivity and stability of the total system rather than those of individual components are emphasized. Modern science and technology is endowed with several elements which can strengthen these linkages and components. On some individual components, considerable work has been done. Research on upland crops and mountain horticulture, as well as coarse grain crops and rangelands in the dry tropical areas, has received considerable attention both at national and international levels (Jodha 1989b). Efforts to impart a farming systems' perspective to R and D have also been made (ICRISAT 1987). Yet, the major gap in the past R and D has been the absence of integrated focus. The latter alone can help diversified and interlinked systems of resource use and production.

Strengthening Resilience: Diversification, flexibility and interlinkages among different production activities, input use practices, and consumption patterns imparted a degree of resilience to the traditional farming systems in the fragile areas. Resilience of the system was also strengthened by factors such as periodic release of pressure through migration or transhumance; a variety of input and product recycling devices; collective sharing systems; and informal regulatory measures to ration the use of fragile resources. Except for a few institutional devices, most of these measures can be strengthened by the new technological components.

To achieve these goals, the focus of R and D will have to be on diversification, flexibility, and local resource-centred interlinked activities. Again, the availability of genetic material of diverse attributes as well as improved knowledge and the capacity to precisely understand interactions between different biophysical variables, offer significant opportunities for the development of options to satisfy these goals (Dregne 1983; Gadgil et al. 1988; Ruttan 1989).

Inter-regional Linkages: Inter-regional linkages, as mentioned earlier, help in sustainability by relaxing internal constraints and facilitating exchange of local surpluses. Under traditional systems, fragile areas had their linkages with other regions largely through harnessing of specific niche and petty trading, unequal exchanges based on large-scale extraction of their resources (e.g. timber from the hills), and periodic migration and transhumance. Such links did help in survival, but are inadequate for sustainability, implying enhanced performance, to meet the increasing demands over time.

The physical and market-based linkages between different regions are a function of the combination of several factors, some of which fall far outside the area of agricultural R and D. However, a basic factor promoting inter-regional exchange is the relative difference in the comparative advantages enjoyed by different regions in specific activities and products. The fragile areas, as mentioned earlier, also have some activities and products that have comparative advantages in relation to other regions. Agricultural R and D can help them by identifying such activities and products and improving their quality and productivity. In the past, this complementarity between fragile regions and other regions did not receive sufficient attention (Jodha 1986b).

CONCLUSION

The prospect of sustainability for agriculture in the fragile areas is severely constrained by the specific features of their natural resource endowments. Sustainability, or rather survivability, in a situation of low pressure on resources was possible through traditional land-extensive practices. In the changed circumstances with high pressure on fragile resources, the required high resource-use intensity (for high productivity) with conservation is not possible through traditional measures. This requires application of modern science and technology blended with the rationale of indigenous practices. Various areas of focus for R and D are indicated to achieve this. Any progress in the suggested direction, however, will depend on the reorientation of agricultural research strategies to suit the specific requirements of these areas. This in turn is largely an institutional rather than a technological problem. The major implication of this conclusion is aptly summed up by R.E. Rhoades (1988), who recommends to the policymakers in mountain areas to start thinking like a mountain. The same applies to other fragile areas. Fragile-area development strategies, to be meaningful and effective, should have a strong fragile-area perspective.

4

Food Security in Fragile Resource Zones[*]

INTRODUCTION

Food security is largely a problem of the poor people and poor areas, who do not have adequate resources to ensure dependable access to required quantity and quality of food through production or exchange. This is more so in areas endowed with fragile and marginal natural resource base, which offers people limited and undependable agricultural production and exchange opportunities for acquiring command over their food requirements (Sen 1981). The bulk of the mountain/hill regions and dry tropical areas in developing countries more readily qualify for this status. In fact (if one excludes the political strife-torn high-food-potential areas), the bulk of the food movement on account of charity is directed to such areas.

A fragile resource is one which cannot tolerate the degree of disturbance implied by the intensity of use associated with a specific activity. Thus, every land resource is fragile, i.e. vulnerable to irreversible damage, when subjected to a degree of use intensity higher than its use capability; and land belonging to use-capability classes IV and above, though good for land-extensive uses, is fragile when assessed with reference to use-intensity associated with prime agricultural lands. One can also describe fragility in terms of low input absorption capacity of resources; limited scope for resource manipulation; and required high level of biochemical subsidization of the natural resource to achieve a level of output comparable to that from the better land resources. Fragility can also be expressed in terms of input–output

*First published as a chapter entitled 'Enhancing Food Security in a Warmer and More Crowded World: Factors and Processes in Fragile Zones' in Thomas E. Downing (ed.), *Climate Change and World Food Security*, pp. 381–419 Berlin: Springer-Verlag & Co., 1995.

ratios, where the fragile lands have higher than average input–output ratios. Described this way, all areas with low potential for crop farming (mountain regions with steep slopes; rain-fed and semi-arid tropical areas with undulating topography, low and unstable rainfall; and coastal areas prone to salinity and waterlogging, etc.) will fall into the category of fragile areas. Despite their apparent differences, for operational purposes, fragility and its associated attributes impart a broad degree of similarity to these areas. This facilitates their consideration as a 'system' for analysing and understanding food security issues. In many of these areas, the threshold limits to maintenance or enhancement of agricultural performance, even by using the inter-regional linkages, seem to have been reached. Further efforts to improve output levels imply over-exploitation of their biophysical resource base and beginning of irreversible degradation (Glantz 1987; Nelson 1988; Grainger 1982; Allan et al. 1988). Their production prospects and output levels, on a per capita basis and in most cases on a per production-unit basis, have declined.

The following discussion, however, is confined to mountain areas, particularly lower and middle mountains in the Himalayas where annual cropping is one of the land-based activities and the rain-fed areas in the arid and semi-arid tropical regions (also referred to as dry tropical areas) in India. Besides their magnitude and the availability of relevant data on them, my close acquaintance with the two regions has dictated this choice. These regions do have substantial pockets where fragility is not a problem, which are excluded from our discussion. Our emphasis is therefore more on specific agro-ecosystems (e.g. fragile, marginal areas) rather than geographic or administrative units, where fragile and non-fragile agro-ecosystems coexist.

To the extent that the essence of food security is the dependable availability of production and exchange opportunities, the fragile resource zones offer a situation where such opportunities could not be lesser. The biophysical constraints and the side-effects of the recent socio-economic developments (partly induced by the former), which will be elaborated in the following sections, tend to obstruct the circumstances that could enhance the range of production and exchange options facilitating food security. Due to fragility and marginality of resources (when compared to prime agricultural lands), and their lower carrying capacities, these areas already manifest overcrowding and its consequences include emerging food insecurity. Most of these areas are already periodically food deficit and account for the bulk of food transfer under charity/relief arrangements. In fact, they exhibit what 'a more crowded world' means and how difficult it is to handle its consequences. This, in turn, may offer some clues for strategies for enhancing food security in the world's areas with food–population imbalances.

In several ways, most of these areas also represent the agro-ecological margins (Parry et al. 1988), where a slight disturbance to their supplies, directly through disruption of their own production systems or indirectly through changes in their external sources of food supplies, due to global warming or other environmental changes, can create a crisis. From the future food security angle, therefore, the fragile areas should be prime candidates for the attention of policy-makers.

Another important factor in discussion of fragile zones is the complex invisibility characterizing them. Being poorly accessible and having marginal status, their problems do not get enough and early attention. Even the factors and processes (often emerging as side-effects of public intervention) accentuating food insecurity in these areas remain invisible (Sanwal 1989; Jodha 1991a). The same applies to the off-site impacts of their poverty, misery and food scarcity. For example, poverty-induced degradation of mountain resources affects the downstream areas; degradation of dry tropics adds to the spread and march of desert to the neighbouring better endowed areas. Problems created by environmental refugees from famine-affected areas are well known.

The conventional 'green revolution' type of approaches to increased food production have not succeeded in most of these areas. Rather, the components of traditional food security and survival strategies and their rationale constitute a reserve of the traditional wisdom which could be profitably incorporated into new approaches to the enhancement of food security.

FOOD INSECURITY AND THE BIOPHYSICAL CONTEXT

Food security (both in terms of accessibility and availability of required quantity and quality of food) in the ultimate sense, is a function of production and exchange options available to the people (Sen 1981). The range and quality of such options in a region are conditioned by the biophysical attributes of its resource base and the pattern of activities (for both production and exchange) it induces or imposes on its inhabitants. Though intra-regional differences in the availability of such options (for different groups and individuals) are an equally important aspect of food security (Kates and Millman 1990), this essay is confined to food security in the context of agro-ecosystem only.

The biophysical context of production and exchange options in the fragile zones may be listed as inaccessibility (more in mountains than in dry tropical plains), fragility, marginality, diversity (or internal heterogeneity, again more in the mountains than in the dry plains), and 'niche' or conditions/products imparting potential for comparative advantages to these regions. While these characteristics are applicable to any region in some measure, their degree is incredibly high and they have a decisive impact on the production performance of these areas.

Inaccessibility: Mountains are inaccessible due to slope, altitude, overall terrain conditions, and periodic, seasonal hazards (e.g. landslides, snow storms, etc.). Dry tropical areas, while less inaccessible, still compare unfavourably with high-productivity, well-watered agricultural areas. The concrete manifestations of inaccessibility are isolation, distance, poor communication, and limited mobility with all their socio-cultural and economic implications. Inaccessibility implies high cost of support systems, impeded exchange linkages, and limited dependability of external support in a semi-closed system. From the food security point of view, these constraints limit the physical production and exchange opportunities. The advantages of scale and technology are restricted. Exchange relations become inequitable. The people have to depend mainly on *in-situ* arrangements for food security, including local resource-centred diversification of food chain, sharing, recycling, etc.

Fragility: Fragility is an attribute of the resource that emanates from the combined operation of slope/altitude or undulating topography as well as geologic, edaphic and biotic factors. Nature's low regenerative capacity and delicate ecological balance are equally important, especially in the dry tropics. Notwithstanding the differences in the relative role of the specific factors in the mountains and the dry tropics, these factors limit the capacity of land resources to withstand even a small degree of disturbance (DESFIL 1988). Vulnerability to irreversible damage due to over-use or rapid changes extends to physical land surface, vegetative resources, and even the delicate economic life-support systems of the dependent communities. Consequently, when resources and environment start deteriorating due to disturbance, they do so rapidly. In most cases, the damage is irreversible or reversible only over a long period (Eckholm 1975; Grainger 1982). Fragility complements inaccessibility (and marginality, to be commented on shortly) in restricting the range and quality of production, exchange and consumption options by preventing resource-use intensification and high productivity. Fragility in a way debars the areas under reference from modern input-intensive technologies to raise food production.

Marginality: Marginality is another characteristic of these areas that is directly related to fragility. A marginal entity (in any context) is one that counts the least with reference to the mainstream situation. This may apply to physical and biological resources or conditions as well as to the people and their sustenance systems. The basic factors contributing to such a status of any area or a community are remoteness and physical isolation, fragile and low-productivity resources, and several man-made handicaps which prevent participation in the mainstream activities (Chambers 1987). The mountains and dry tropical plains being largely marginal areas, compared to the prime agricultural regions, share the above attributes of marginal entities and bear the consequences of such a status in different ways (Bjonness 1983; Gadgil et al. 1988). Marginality shares with fragility a number of implications. Besides constraining the range and quality of production options, it reduces the people's capacity to undertake and benefit from high-cost, high-productivity opportunities. Due to marginality, the inter-system linkages make the terms of exchange unfavourable to these areas and their people; the mainstream decision-makers ignore them and treat them as national liabilities.

Diversity or Heterogeneity: Fragile areas are internally heterogeneous even in terms of degree and nature of fragility or marginality. The mountain areas display immense variations among and within eco-zones, even over short distances. This heterogeneity is a function of interactions between different factors ranging from elevation and altitude to geologic and edaphic conditions. The diversity in arid and semi-arid tropical plains is primarily because of topography, soils and precipitation differences. Water, a homogenizing factor, being limited, the diversity is accentuated. The biological adaptations (Dahlberg 1987) and socio-economic responses to these diversities (Nogaard 1984) also become heterogeneous consequently. Diversity becomes both a constraint as also opportunity. As a positive attribute, by supporting interlinked activity patterns, diversity can help enhance food security in these areas. At the same time, diversity implies greater location specificity of production activities, which lessens the gains from specialization and scale of activity. It also restricts the applicability of generalized (technological and institutional) measures (Jodha 1991a).

'Niche' or Comparative Advantage: Owing to their specific environmental and resource-related features, both mountains and dry tropics provide a 'niche' for specific activities or products (Banskota and Jodha 1992b; Jodha 1989b). For mountain areas the examples may include: specific valleys serving as habitats for specific medicinal plants; mountains acting as a source of important high-value products (e.g. fruits, flowers, etc.); and mountains serving as sources for hydropower production. The dry tropical plains may have comparative advantages for land-extensive activities (e.g. pasture-based animal husbandry), highly moisture (or humidity)-sensitive cultivars, such as some coarse grains, etc. The local communities make use of these 'niches' through their diversified activities. There is scope, however, for better harnessing of niche, which requires both infrastructural facilities and financial as well as technological resources, which the local inhabitants lack. Proper harnessing of niche can help in food security through directly usable products or tradable high-value products, but their reckless exploitation for the benefit of mainstream economy can result in the destruction of niche (Jodha 1991b).

To sum up, constraining food security in the fragile zones are the limited physical possibilities, the limited relevance and transferability of technologies from outside and difficulties of evolving them internally, and the lack of infrastructural support. The resource characteristics of fragile zones also obstruct the exchange-related options that could enhance people's entitlement to food resources. Not only is the scope for surplus production and processing limited, but market and infrastructural support to encourage exchange activities are weak and too costly to create. Chart 4.1 sums up these characterisitcs and their implications for food security.

On account of their biophysical specificities, the fragile zones are unable to integrate themselves as equals with the mainstream economy. Development programmes and processes bypass them. Rather, through various interventions irrelevant measures are imposed on these areas, which lead to the disintegration of their production systems and people's survival strategies (Jodha 1991a, 1991b; Whitaker et al. 1991). This will be elaborated later.

FOOD SECURITY AND THE SOCIO-ECONOMIC CONTEXT

Despite the harshness of the terrain, substantial human populations have survived and multiplied in the fragile zones. At a relatively low level of demand, they managed their food security through a two-way adaptation process, (a) by adapting the needs and demand (and mechanisms to satisfy them) to the constraints and opportunities offered by the natural resource base, and (b) as far as permitted by their skills and capacities, amending the resource base (e.g. through terracing, water harvesting, etc.) to satisfy the communities' growing needs. A glance at the traditional farming system and resource management practices will corroborate this (Guillet 1983; Whiteman 1988, Jodha 1991a; Jodha and Partap 1993; see Chart 4.1).

The key dimensions of traditional food security arrangements, manifested in different practices, include the following:

a. Diversification and flexibility of both food production and consumption activities, making fuller use of the temporal and spatial variability of the environment and production resources.

b. Local resource regeneration and recycling of products and inputs to meet the constraints imposed by fragility, marginality and inaccessibility characterizing such areas.

c. Demand management through informal resource and product rationing and collective-sharing arrangements involving various forms of group action, petty exchanges, transhumance, etc.

d. Resource upgrading and new crop introduction as permitted by the slowly evolving folk agronomic knowledge.

The literature on ethno-sciences, traditional farming systems, indigenous knowledge systems and traditional forms of rural cooperation, which is getting increasing attention in the recent years, can furnish several examples of how people in fragile zones attempt protection and multiplication of production/consumption options within the constraining biophysical environment. A ready inventory of such measures is offered by the semi-popular, user-focused journals/newsletters, such as *Indigenous Knowledge and Development Monitor* (The Hague), *ILEIA Newsletter* (Leusden) *IIED Gatekeeper Series* (London), *Honey-Bee* (Ahmedabad), and *Forests, Trees and People Newsletter* (Uppsala).

A few important preconditions associated with the traditional food security measures in the fragile areas included the following. (a) They involved adaptation to a high-risk, low-productivity environment. (b) The choice and combination of food security measures required empathy for the local environment. (c) Their implementation needed strict adherence to formal/informal institutional norms to ensure resource use regulation and group action. (d) Their feasibility and efficacy was closely associated with low pressure on land resources, which permitted land-extensive production practices and acceptance of stable but low production (as in the case of subsistence-oriented systems). Of late, however, the circumstances supporting the effective use of traditional measures for maintenance and multiplication of food security options have radically changed, reducing feasibility and efficacy. The changes are a consequence of rapid population growth, increased extent of state intervention and increased role of market forces.

Population Growth: Since the 1950s, the population of most of the areas has almost doubled. For instance, the population density in the hills and mountain areas of Nepal in 1991 was 26 persons/km^2. The corresponding figures for areas covered by the Hindu Kush–Himalayan region in India and Pakistan were 73 and 56 respectively. The Chinese parts of the Himalayan region had a population density (mainly due to sparsely populated Tibet) of 12 persons/km^2 for the year 1987 (Sharma and Partap 1993). In the dry tropical areas of India the subdivision-level data (for 1981) for eight arid or semi-arid states showed a population ranging from 67 persons/km^2 in more drier districts to over 300 persons/km^2 in better rainfall districts (Jodha 1986a). The bulk of the population in these regions directly depends on agriculture and already represents an overcrowding of these areas.

The increased population has adversely affected food security in two ways. First is the obvious decline in per capita availability of food. Despite the extension of annual cropping to steep slopes and sub-marginal lands in the mountains and the dry tropical areas, per capita availability of cultivated land has declined—from 0.17 ha. in 1971 to 0.13 ha. in 1981 in Nepal, from 0.19 ha. to 0.13 ha. in Himachal Pradesh, from 0.30 ha. to 0.16 ha. in the UP hill districts (Sharma and Partap 1993). The per capita cultivated land in the subdivisions of the dry tropical states of India declined from 0.85 ha. in the 1950s to 0.41 ha. in the 1980s (Jodha 1986a). Combined with this fall in land availability was the fall in land productivity. In the hills and mountains of Nepal the per hectare average yields of paddy, maize,

Chart 4.1. Key Elements of Traditional Food Security Arrangements in Fragile Resource Zones

Conditions Constraining Production/ Exchange Opportunities	Traditional Measures to Enhance Food Security (Production/ Exchange) Options
Inaccessibility Isolation, semi-closedness, poor mobility; Limited access to and dependability of external support (inputs, products); High cost of mobility, infrastructure, support systems and production/exchange activities Detrimental to harnessing of niche and gains from trade.	Local resource centred, diversity of production and construction activities; Local resource regeneration and resource use regulation/ rationing (by informal institutional means); Collective sharing, recycling of resources/products, limited external input use; Nature and scale of operations as permitted by degree of mobility and local resource availability (e.g. transhumance).
Fragility Resources highly vulnerable to rapid degradation, unsuited to high intensity/productivity uses; Limited, low productivity, high-risk production options; little surplus generation; High overhead cost of resource use, obstacles to infrastructure development, under-investment; People's low resource capacity preventing use of high-cost, high-productivity options.	Resource upgrading; focus on low-intensity uses; Resource protection, usage regulation; Diversification involving a mix of high- and low-intensity land uses; a mix of production and conservation measures; Local resource regeneration, recycling, limited external input use; dependence on nature's regenerative processes and collective measures.
Marginality Low productivity, limited and risky production options; little surplus generation and reinvestment; Resource scarcities, inability of people to benefit from high-cost, high-productivity technologies/options.	Focus on security of subsistence needs, low and local input use; integration of production/ consumption activities; Diversification, recycling, collective sharing; greater dependence on nature's processes, self-provisioning.

(Continued)

Chart 4.1. continued

Conditions Constraining Production/ Exchange Opportunities	Traditional Measures to Enhance Food Security (Production/ Exchange) Options
Diversity Heterogeneity-induced strong location specificity of production options; Limited usee of options involving wider applicability and benefits associated with scale.	Small-scale, interlinked diversified production/consumption activities; temporally, spatially differentiated for fuller use of environment; Location-specific, integrated multiple activities; stability of total system.
Niche Unique and complex potential opportunites with comparative advantage; Opportunities for production and exchange remain under-utilized for want of resource, infrastructure.	Specific opportunities harnessed for petty trade and self-consumption; Niche-harnessing integral part of diversified resource use.

Sources: Based on evidence and inferences from Whiteman (1988), Sanwal (1989), Pant (1935), Jochim (1981), Guillet (1983), Bjonness (1983), Jodha (1990b, 1991a, 1991b, 1989b, 1980, 1975), Gadgil et al. (1988), Jodha and Singh (1990), Jodha and Partap (1993); Walker and Ryan (1990), Singh (1992).

millet and barley declined by 19 per cent, 21 per cent, 18 per cent and 9 per cent respectively during the period 1980–7 as compared with the preceding decade (K. Banskota 1992). In the Kumaon hills (UP, India), decline in the yield of maize, millet and oilseeds has also been reported (Swarup 1991). According to village-level investigations under ICIMOD-sponsored studies, most of the mountain villages in Nepal and India showed decline in yields of important food crops. For the dry tropical region, the areas that have not been able to adopt high-yielding varieties of sorghum, millet, etc. have shown decline in yield (Jodha and Singh 1982).

The second and even more important adverse impact of population pressure is that the measures and mechanisms directed at option maintenance/enhancement (through diversification, local-resource regeneration, collective sharing, recycling, etc.) have become unfeasible or ineffective. Most of the village-level studies conducted by ICIMOD,

Chart 4.2. The Changing Complex and Orientation and Components of Food Security Strategies in the Fragile Resource Zones

Traditional Strategies	Present-day Strategies	Implications of Present-day Strategies
Orientation/context		
Community/household-operated measures, part of subsistence-oriented, self provisioning systems evolved in keeping with the local circumstances; guided by informal institutional arrangements and state of folk knowledge, focus on stability and productivity of total systems	State administered/patronized/ encouraged activities as a part of formal, generalized development/ welfare programmes: focused on infrastructure, agricultural production programmes, distribution network, public sector investment, large-scale projects to raise employment, income, productivity, etc., heavy reliance on external experiences	'Universalization' of food security issues, relaxation of local situation based food constraints; itegration with the main-stream situation. Marginalization of local initatives and measures; misamatch between imperatives of resource characteristics and the features of generalized interventions with negative side-effects affecting food security.
Resource focus		
Diversified resource use; balancing; extensive–intensive land uses, production conservation needs; focus on local resource potential, resource rationing, recycling and regeneration, upgrading, local resource control	Priority to production over conserva tion, high use-intensity, over-extraction, large-scale resource trans-formation e.g., by irrigation projects (ignoring location specificities, diversities)	Over-exploitation, degradation of resource base, sustained and subsidized by the state; bypassing local concerns; emerging long-term unsustainability.
Agricultural production front		
Diversified, interlinked multiple land-based activities; practices driven by local demand, local resource and local	Indiscriminate intensification, narrow specialization, selective commercializ-ation; sectoral segregation of land-	Rapid resource depletion, increasing bio-chemical and economic subsidization of production systems; emergence of dual

knowledge with focus on recycling, flexibility, and stability; productivity of total system; integration of production and consumption activities	based activities; focus on high cost, high productivity; external inputs and technologies, subsidised by the State	(transformed and stagnant) sectors, imposition of inappropriate external options; increased dependency
Demand management Supply-driven demand situation; focus on (diversified) food chain rather than on food items, informal demand rationing, sharing and recycling practices, transhumance, etc.	Focus on supply side mainly, dependence on external supplies and relief, focus on selected foodgrains; increased external demand for selected local items, unregulated market pressures	Rapid growth of pressure on local resources and their degradation, permanent dependency on relief, marginalisation of collective sharing, demand rationing, food chain, etc.
Linkage and Exchange Strong linkages between temporally/ spatially differentiated activities (e.g., farming–forestry linkages); inter-system (i.e. external) linkages limited; need-based and local capacity driven harnessing of niche for petty trading; transhumance and migration	Improved accessibility-led physical and market integration, better access, external market-driven over extraction of niche and exchange at unequal terms, under-valuation of products, limited plough back of resources	Selective over-extraction of resources with limited local benefits, development of dual sector economy; people's alienation from resources; domination of mainstream decision makers, disregard of multiple small niche

Sources: Based on evidence and inferences from Banskota amd Jodha (1992a, 1992b), Jodha (1993, 1992, 1991b, 1986b), Gadgil et al. (1988); and Whitaker et al. (1991).

ICRISAT and others indicate that the extent of intercropping, the extent of crop rotations, the number of crops and combinations and the forms of diversification and linkages have declined over time. Decline of such practices means resource depletion and reduced carrying capacity, finally affecting food security. A crucial element of these practices is a low man-to-land ratio. When that is reversed, attitudes change, discounting group action and adherence to social sanction for collective arrangements. Chart 4.2 gives a broad idea of these changes and their inadequacies.

Integration and Intervention: Partly the attitudinal change is a product of increased state and market intervention, which is directed at the process of modernization, development, etc. The increased physical and market integration with the mainstream economy, the state's taking the responsibility for food security in these areas as a part of the welfare-cum-development programmes and the focused intervention to raise food production and supplies, have certainly helped in relaxing the biophysical constraints of the fragile regions, but most of them, by not being sensitive enough to the specific characteristics of such regions have also generated several negative side-effects.

The rapid erosion of land-extensive traditional production practices due to demographic changes was followed by a phase where the search for food security implied high use-intensity of fragile land resources. This phase coincided with increased state intervention in socioeconomic affairs of the rural communities (Jodha 1991b, 1992). The intensification, ignoring resource limitations, took place at both extensive and intensive margins. This has imparted several unique features to the food security strategy not found in the traditional arrangements. Chart 4.2 summarizes a few of them. The changed resource use practices followed by the farmer and encouraged by development intervention, including through subsidization, have led to the emergence of several negative trends, often described as indicators of unsustainability (see Chapter 3).

MISGUIDED GOOD INTENTIONS

Some of the negative features of well-intentioned government food security policies for the fragile regions are pinpointed in the following paragraphs.

'Universalization' of Food Security Issues: Food security in the new dispensation is focused through increased employment and income opportunities (i.e. entitlement enhancement) to be generated by

generalized development projects, agricultural production prog-rammes and public distribution systems based on mobilization of food from the surplus producing areas (Drèze and Sen 1990). This has definitely reduced the role of local constraints (including limited food production possibilities). Despite lapses and leakages, the state-sponsored arrangements have drastically reduced the occurrence of famine situations. Nevertheless, public interventions have generally ignored the location specificities and diversities associated with both the demand and supply (production) aspects of food security in the fragile and marginal areas. The food under the new arrangements means a few cereals rather than a chain of seasonal, spatial, ethnic and agro-climatic context-wise differentiated items. Retrospective surveys using case history method in the hills of Nepal and UP (India) recorded more than twenty food items traditionally consumed by the villagers. Now this number is reduced to around five (Singh 1992). The comparable details of food items recorded in the dry tropics of India ranged between ten and five. These items excluded several minor items (Jodha 1985a). Today, access to food is largely deter-mined by an individual's capacity to acquire it (i.e. entitlements), with little collective-sharing arrangement. In the dry tropics forms of sharing arrangements during droughts declined from twelve during 1963–4 to barely three in 1980–1 (Jodha 1975, Chapter 7). Further-more, elements such as informal rationing and recycling, etc. have no place under the new arrangements. More importantly, the 'entitlement enhancement' approach also had a mixed success, as the development intervention (again due to their generalized approach unsuited to the specificities of fragile resource zones) could not make a significant dent on the situation except in selected (relatively better endowed, accessible and patronized) pockets. Also, due to the mismatch between the features of resource base and the attributes of interventions, the latter had serious negative side-effects, which eroded the range of production options traditionally available to the people.

Indiscriminate Resource-use Intensification: The indiscriminate resource-use intensification under the new arrangements is clearly visible most in the agricultural production programmes, focusing on food self-sufficiency, agricultural intensification and narrow specializ-ation, and sectoral segregation.

While fully sympathizing with the policy-makers' concern to feed the increasing population, one may question the undue focus on food self-sufficiency in the fragile areas, especially when food is defined in a narrow foodgrain sense and isolated from the total food chain (a range of seasonally and spatially variable consumable items) generated by the diversified and interlinked land-based activities. This disregards the historical experience that even with relatively smaller populations most of these areas had rarely been self-sufficient in food (defined narrowly). The deliberate control of demand, recycling and collective sharing, periodic migration, focus on petty trading involving barter exchange of high-value products for food, strengthening of the totality of the food chain, etc. were the means of adjusting to food scarcity in the past.

The new strategy, focused on foodgrain rather than the total food chain and ignoring the regions' petty but multiple niche (i.e. production with comparative advantage), missed the essence of food security in the fragile resource context. By implication, they also disregarded the value of various features of the traditional measures to manage food systems.

Furthermore, the foodgrain-focused strategies led to the extension of cropping to sub-marginal lands (and steep slopes) not suited for annual cropping, and increased the cropping intensity on traditional crop-lands. This amounted to reduced extent of diversification and interlinkages of different land-based activities, reduced local resource regeneration and associated flexibility, physical degradation of fragile lands, increased biochemical and economic subsidization of the agricultural production systems and reduced average yield of different crops due to inclusion of marginal areas into croplands.

Agricultural intensification applied to both food crops as well as high-value cash crops such as vegetables and fruits in mountains and oilseed crops in dry areas. Agricultural intensification had two inter-related key dimensions. First, narrowed focus of programmes in terms of choice of priority crops or their varieties as well as their emphasized attributes (e.g. growth of grain yield as against stability of total biomass). Second, input-intensive cropping, involving increased application of high-cost external inputs. Both were merely transplants of the agricultural experience from prime lands (Jodha 1991a, 1986b; Jodha and Partap 1993; V. Singh 1992; Whitaker et al. 1991).

Apart from high dependence on external inputs, this led to the marginalization of several important local crops and varieties. For instance, in several valleys of the middle mountains of Nepal and the UP hills in India, there used to be more than a dozen types of rice. Today, in most cases, one finds hardly two or three varieties developed by agricultural research centres, research stations, etc., that occupy the bulk of the rice-growing areas. The picture is similar for horticultural products in many parts of the Hindu Kush–Himalayan region, and millet and legume varieties in the dry tropics (V. Singh 1992; Jodha 1986b).

Input-intensive cropping unavoidably follows the focus on yield levels and the choice of technologies that facilitate it. Both for multiple cropping and for high yields of individual crops, the dependence on external (chemical) inputs has rapidly increased. This is partly due to the very design of the technologies employed and partly due to the inability of the local resource-regenerative systems (e.g. nutrition cycling through farming–forestry linkages, biological control of diseases and pests through specific crop rotations, etc.) to meet the high and temporally concentrated demands of the new systems.

The required diversity and interlinkages of activities in the fragile resource zones are ignored not only in the production programmes at field level but more so at the policy and planning levels. To suit the existing system of division of labour (or rather division of bureaucratic power), agricultural activities (relating to cropping, animal husbandry, horticulture, forestry, etc.) are disintegrated in terms of departments, line agencies, etc. Development activities, including project formulation, investment allocation, foreign assistance, etc. are separated accordingly. This ill accords with the ground reality where different resource characteristics are interlinked (e.g. fragility and diversity) and different land-based activities are interdependent (e.g. cropping and animal husbandry). The sectoral segregation of decision-making and resource allocation not only makes development interventions ineffective but often counter-productive (Sanwal 1989).

A further lacuna of state-patronized food arrangement is that it is not particularly known for management of the demand side, the emphasis being on the supply side, be it focused crop production programmes, raising food purchasing power of the people, or subsidized public distribution systems. The demand aspect, except during the occasional cases of formal food rationing, is rarely emphasized. Not only elements of traditional arrangements such as diversification of food requirements, recycling, sharing, deliberate consumption curtailment, etc. are disregarded, but the state policies and programmes deliberately help raise the pressure of external market demand on resources and selected products of fragile resource areas. Such products relate to the 'niche' of fragile resource zones such as timber, vegetables and fruits in the mountains and animal products, oilseeds, and pulses in the dry tropical areas. Besides the over-exploitation of fragile resources, an over-emphasis on these products has backlash effects on other food crops, which are an important part of the local food system but have little demand in the external market. Displacement of maize and minor millets by high-value vegetable crops in the mountain areas and the displacement of millets by oilseeds in the dry tropics are cases in point. Ironically, due to marginality-related attributes, producers of the high-value crops often do not receive favourable prices. In addition, there are no mechanisms to regulate the external demand pressure through fiscal or other means. The consequence is further use intensification and degradation of the natural resource base of fragile areas.

An important dimension of demand management relates to external market linkages and terms of exchange. The improved accessibility-led physical and market integration of fragile areas with the mainstream urban economies has definitely reduced the vulnerability of these areas to the severe food scarcities and famines (Drèze and Sen 1990). At the same time, the integration of unequals does not necessarily help the weaker and marginal entity. Due to unequal terms of exchange and under-pricing of their products, integration has more often led to draining out resources from these regions rather than bringing them in. This is reflected in product pricing as well as compensation norms for resource extraction from these areas. Examples are minerals, timber, hydropower, animal products, herbs, fruits, etc. Such unequal terms of trade result in the local people's loss of control over their resources; inadequate local gains of harnessing the niche (to raise their food purchasing power); and marginalization or disappearance of several interlinked production activities with food components.

An extreme case of unequal exchange is provided by the following. The state or its agencies often acquire whole land tracts for harnessing specific products (e.g. minerals) or for preserving the biodiversity and wildlife (as under parks and sanctuaries), or for defence-related activities. This phenomenon has high incidence in fragile resource zones because, despite rapid degradation of their resources, these areas are the only left-over habitats of multiple fauna and flora, they are the major areas with relatively sparse and voiceless population, and they continue to bear the consequences of their marginal status.

Yet another dimension of such unequal inter-system linkages is their side-effect on the organic and economic integrity of the traditional production systems. Nationalization of forests or state acquisition of common property pastures, adversely affecting the farming–forestry–livestock linkages that supported the local food systems, is one example (Yadav 1992; Jodha 1992). The direct links of modern dairying with the urban market and substitution of some crucial local inputs by external inputs are other examples where the integrity of the total production system has broken down.

The consequences of the breakdown of resource-regenerative diversified production systems and indiscriminate resource intensification (under the pressure of population growth, enhanced market forces, development interventions, etc.), are manifested in various forms of resource degradation, falling resource availability and resource productivity, and the degree of desperation in people's responses to the deteriorating situation (Ives and Messerli 1989; Jodha 1992).

Based on the observations and quantified evidence at the micro-level, from different areas within the fragile resource zones under review, a few of the indicators of negative changes are summarized in Chart 4.3. These negative changes, covering a period of just four to five decades, manifest the extent of the emerging crisis and scarcities in the fragile zones. Viewed differently, they represent what is described as cumulative type (as against systemic type) of global environmental changes.

FOOD SECURITY AND ENVIRONMENTAL CHANGE

Broadly speaking, systemic change is one which while happening in one locale can affect changes in the system elsewhere. The underlying activity need not be widespread in scale, but its potential impact is global. CO_2 emissions from limited activities, having impact on the great geosphere-biosphere systems of the earth and causing global warming are a prime example. The cumulative type of change refers to localized but widely replicated activities, where a change in one place does not affect change in other places. When accumulated, they may acquire a scale and potential that can influence the total global situation in specific ways. Widespread deforestation and extractive land-use systems and their potential impacts on global environment serve as examples. Both types of change are the products of nature–man interactions and they are linked to each other in several ways (Turner et al. 1992).

In scientific discourse and policy advocacy, 'systemic changes' and associated 'geocentric perspectives' are the primary focus. The variables

Chart 4.3. Indicators of Emerging Crisis and Scarcities Following the
Increased Pressure and Inappropriate Resource Intensification
in Fragile Resource Zones

Mountain Areas	Dry Tropical Areas

Resource Degradation

Soil: Increased landslides, mud-slides, collapse of terraces, rapid top soil disappearance from steep slopes

Soil: Increased salinity, erosion of top soil shifting sand covering fertile lands

Water: Reduced (dry season) water flows for irrigation and domestic use; frequent flash floods; increased runoff and soil erosion during rainy season

Water: Ground water salinization, deepening of water-tables; increased moisture insecurity and frequency/severity of drought; decline and unfeasibility of water harvesting systems

Vegetation: Decline of forest/pasture area and productivity; disappearance of minor plant species; biodegradation; reduced biodiversity and biomass availability

Vegetation: Emerging plantlessness in forest/pasture, reduced perennial, increased inferior annuals, thorny bushes; reduced carrying capacity and diversity

Resource Availability

Per capita decline in cropland, irrigated land, grazing land, common lands; reduced access/availability of forest and its products (fuel, fodder, fiber, food items)

Decline in per capita cropland, per animal grazing land, per hectare local resource input (e.g. manure); shortage and instability of natural biomass; increased landlessness and reduced access to water facilities

Resource Management/Productivity

Decline in agricultural diversity, farming-foresty linkages; decline in yield of several crops, infeasibility of traditional land extensive resource management practices and local resource regeneration, recycling; high bio-chemical subsidization of the production systems

Decline in local resource regneration/recyling, diversity, flexibility in agriculture; decline in crop/animal productivity, collective resource management and risk sharing, disintegration of traditional farming systems (intercropping, crop rotations, etc.)

Impacts/Responses

Desperation and choice of inferior options: Abandonment of old terraces; cropping on more steep slopes; substitution of shallow-rooted crops for deep-rooted

Desperation and choice of inferior options: Limited abandonment of saline, eroded soils; using land despite unstable and reduced productivity; 'mining the

(Continued)

Chart 4.3 continued

Mountain Areas	Dry Tropical Areas
crops on eroded slopes; substitution of small ruminants for cattle due to lower carrying capacity of pastures; increased seasonal migration of men and animals, marginalization of collective sharing; high external resource dependency and indebtedness	resources', e.g. groundwater; extension of cropping to submarginal lands, premature harvesting/lopping of trees, overgrazing; increased dependency on relief and external resources; decline of collective self-help; substitution of sheep/goat for cattle due to reduced area/yield of pastures; migration, destitution

Sources: Based on a synthesis of findings of different studies. For mountain areas see Jodha (1990b, 1991a), Shrestha (1992), Singh (1992), Shutain and Chunru (1989), Hussain and Erenstein (1992). For dry tropics Jodha (1986a, b, 1989a, b, 1990a, 1991a, 1993), Gadgil et al. (1988), Jodha and Singh (1982, 1990).

and processes covered by them are more complex and difficult to analyse at the current level of scientific skill and data availability. The uncertainties associated with the change scenarios based on them are, therefore, not unexpected. The 'cumulative type' of change and associated 'anthropocentric perspectives', on the other hand, present a situation much simpler and involving less uncertainties and unknowns. It is much easier to relate them to the human approach to constraints and opportunities presented by the environment. Yet, since they are not sufficiently emphasized in global-change related work, their potential for guiding response strategies remains underutilized (Kasperson et al. 1990). Similarly, the complementarities between the two types of change (both in terms of causes and consequences) are not fully understood and analysed. This is largely because of the skewed perspectives of the work and debate on the global environmental change (Jodha 1993). Chart 4.4 presents some details indicating the skewed nature of the perspectives on global change and their implications.

The vegetation-related impacts of global warming (though involving significantly different degrees and processes in the two agro-ecosystems) can accentuate the already visible trends in deforestation, reduced bio-diversity and biomass supplies and the dependent, diversified farming systems, with direct and indirect consequences on the food situation. Most importantly, in both mountains and the dry

Chart 4.4. Indicators of the Skewed Perspectives on Global Environmental Change

Elements Prominently Focused	Elements Under-emphasized
'Systemic' type of change: Focus on biogeochemical variables and their interaction process relating to the functions and operation of the geosphere biosphere systems of the earth	*'Cumulative' type of change:* Localized and widely replicated changes in different variables and processes of resource use (when accumulated) influence the global systems
'Geocentric perspective' Focus physical dimensions in the natural science framework; concentration on geobiological variables and their complex interaction patterns, with little direct incorporation of the human dimension of changes and change-processes	*'Anthropocentric perspective'* Primacy of nature–society interactions with focus on their importance to the society, potential mechanism for understanding and handling cumulative change with some possibility of influencing the impacts of 'systemic changes'
Other associated aspects Emphasis on long time horizon (decades/centuries) and inter-generational issues; focus on terminal impacts involving selected variables (e.g. sea level, temperatures rise, shift of climatic zones, etc.) affecting fundamental equilibrium of world system and atmosphere; analytical methods and material used involve high degree of complexity and sophistication, limited transparency, information on several unknowns, multiple uncertainties, conjectural nature of predictions	*Other associated aspects* Sensitivity to both intra-generational and inter-generational issues; analytical approaches simpler and oriented towards integration of change processes in the current problem-solving mode; predictions, action/advocacy focus on short or medium planning horizon, greater ease and possibility of associating causes, consequences of responses to change; greater possibility of integrating geocentric and anthropocentric perspectives
Advocacy and action High 'scare and noise' potential of issues covered, (e.g. doomsday predictions); approaches to abate/adapt to changes: obstructed by uncertainty of change scenarios, induce higher discounting of the potential options, inject vagueness about gains and sacrifices and create more panic and debate than concrete action	*Advocacy and action* Possibility of evolving options within the received (and modified) framework of handling current crisis situations in local contexts, greater scope for clearly associating cost and benefits, greater certainty of potential options and their easy acceptability to decision makers; possibility of dual purpose options to handle current and future 'impacts'

Source: Adapted from Jodha (1993).

tropics, the warming-led impacts can reduce the resource users' sensitivity to the imperatives of features such as fragility, marginality, diversity, etc. Materialization of such possibilities will only accentuate the current consequences of neglect of the above imperatives. On the basis of the projections by various general circulation models (GCMs) relating to these areas, important details can be summarized (see Chart 4.5).

If the positive impacts of warming (e.g. possibility of better cropping prospects through improved moisture and CO_2 fertilization impacts, etc.) are excluded, the key inferences from Chart 4.5 suggest that global warming can potentially influence the production environment and the resource base in mountain areas by accentuating deforestation and changing the precipitation (including snow melt) pattern. It can result in dislocation of production systems and create more incentives and compulsion to over-extract and degrade the resource base, finally affecting the food situation. In the dry tropics, changed precipitation pattern without the certainty of gains in terms of longer growing season may impart uncertainty to the future. More importantly, there is a possibility of the present marginal areas being exposed to greater weather risks and drought-induced scarcities (Parry et al. 1988; Downing et al. 1996).

More importantly, resource charactistics such as fragility, marginality and diversity in the fragile regions can become accentuated through the changed biophysical make-up of the regional environment following the impacts of warming (e.g. increased soil salinity due to increased evaporation in dry areas; vulnerability to rapid gully erosion on mountain slopes due to high rains, etc.). Again, warming can create circumstances in these regions that could induce or compel people to further disregard the imperatives of resource features such as fragility, marginality and diversity with all the negative consequences already discussed.

Chart 4.6 presents the broad indicators of possible worsening of the current crisis with the materialization of potential impacts of warming in the context of fragile resource zones. The current negative trends relating to the production base, resource productivity and management practices, and the people's vulnerability to these changes have several elements which strongly converge with the components of what are described first-, second-, and third-order negative impacts of global warming. For instance, precipitation-based dislocations following warming can accentuate the present problems of soil erosion, disturb the water harvesting systems, and the well-adapted farming systems

Chart 4.5. Possible Impacts of Warming in the Fragile Resource Zones

Hindu Kush–Himalaya Mountain Regions	Dry Tropical Regions in India
(a) *Forests* (Unmanaged eco-systems): Qualitative/quantitative changes, disappearance or spatial movement of species; reduced biodiversity, increased deforestation and dislocation of biophysical/hydrological flows and dependent life-support systems.	(a) *Precipitation (mixed picture):* Temperature increase; increased precipitation from convective rainfall with higher intensity; high rainfall, high runoff, high soil erosion (due to plantlessness and undulating topography), high evaporation with uncertain prospects for longer growing seasons and higher crop production.
(b) *Precipitation:* Higher rainfall (convective high intensity rains), increased runoff, flash floods, soil erosion, mud/landslides, influencing the biophysical functions of nature and the farming systems.	(b) *Areas on margin:* Area currently frequented by prolonged droughts and crop failures may further suffer with greater variability of rainfall, exposure to greater uncertainty and weather risk.
(c) *Seasonal snow melt changes:* Disturbance to hydrological cycles and seasonality of waterflows and well-adapted resource use/production systems, diversity, etc.	(c) *Impact of a and b:* Disturbance of well-adapted farming systems, reduced ground water recharge, increased soil salinity, reduced regenerative capacity of nature to support crop and natural biomass-based mixed farming and people's food systems especially in marginal areas.
(d) *Resource characteristics:* Fragility, diversity, niche, etc. and their interrelationships to be affected by a, b and c; new constraints (compulsions) and new opportunities (inducements) for higher resource extraction ignoring fragility, marginality, etc.	(d) *Resource characteristics:* Fragility, marginality, etc. may accentuate and adversely affect the resource use systems; increased dependency on external support, food insecurity.

Note: The table excludes positive impacts of warming as they hopefully will not adversely affect the food security situation.

Source: Table based on Jodha 1989c and 1993 and Topping et al. 1990.

in both mountain areas and dry tropics. Such changes will accentuate the already visible negative changes relating to these variables.

ENHANCING FOOD SECURITY: REVERSING THE NEGATIVE PROCESSES

Applied to food security, the above reasoning implies that enhancement of food security in the context of increased population and accentuated resource degradation processes (i.e. cumulative type of environmental changes) should also serve as a strategy against food insecurity due to impacts of global warming. Such dual-purpose strategy should focus on arresting and reversing the currently visible negative trends relating to resource availability, resource productivity, and people's access to opportunities based on harnessing the regional resources, etc. In practical terms, it would mean avoiding the mistakes of the past and building on the positive experiences of history (Kates and Millman 1990). The focus of new strategies has to be on reversing the processes that contributed to the currently visible negative trends. To do so, the involved issues should be addressed at different levels ranging from macro-level policy aspects to the field-level operational steps.

At the policy/planning level, the recognition and incorporation of the following interrelated issues is a prerequisite for initiating the processes that may help any steps directed towards enhancing food security in the fragile resource areas.

(a) Understanding and internalization of the imperatives of the specific resource characteristics (fragility, diversity, niche, etc.) is essential to reorient public interventions to suit the situation of fragile resource zones. This will help amend and evolve interventions in keeping with the specific features of these areas, and reduce the extent and imposition of generalized (less relevant) approaches evolved in other contexts.

(b) Recognition and application of the rationale of traditional food supply/demand management measures to add relevance and effectiveness to the present-day steps adopted for food security. This will mean reorienting several production and resource use strategies and decentralization of activities.

(c) Learning from experiences of unsuccessful development intervention and its reorientation to suit the specific circumstances of fragile resource zones. This also implies amendments in

Chart 4.6. Possible Accentuating Impacts of Global Warming on Current Indicators of Crisis/Food Insecurity in Fragile Zones

Potential Impacts in Mountain Areas (M), and the Dry Tropics (D)	Indicators of Emerging Crisis (Chart 4.3) Mountain areas	Likely to be Reinforced in Dry Tropical Areas
Precipitation-based dislocations Increased runoff, floods, seasonal changes (M); soil erosion (M, D), landslide/mudslide (M); increased weather risks (D); disruption of adapted production systems, decline in supplies (M, D)	Increased land/mudslides, flash floods; off-season water scarcity; collapse/abandoning of terraces; decline of cropland, irrigation facility; reduced cropping diversity, productivity; seasonal food deficit; etc.	Increased soil and ground water salinity, deeping of ground water, decline of water harvesting systems, icreased frequency and severity of droughts, decline in crop/animal yields, increased external dependency
Forest, biomass-related changes Deforestation, reduced biodiversity (M); reduced biomass productivity (M, D); instability of dependent production systems, resource regeneration/stability (M, D)	Deforestation, bio-degradation, decline of farming–foresty linkages and resource regenerative processes; reduced biomass; decline of diversification, reduced resource and product availability	Increasing plantlessness, shifting sands, reduced biomass supplies, increased animal migration, decline of mixed farming systems, reduced resource regeneration, various forms of desertification
Pressure for disregard of resource specificities Resource intensification, resource degradation, decline of adapted production systems, practices	Decline of well-adapted production systems; focus on indiscriminate resource intensification, insensitivity to resource limitations; over-exploitation of niche; desperation in resource use	Indiscriminate resource intensification, disregard of resource limitations; decline of diversifications, recyling, regeneration; desperation in resource use, poverty resource degradation cycle

Note: For more details see Jodha 1989c and 1993.

generalized approaches and measure to suit the needs and capacities of fragile areas.

(d) Learning from the experiences of success stories within the overall environment characterized by unsuccessful initiatives. In fact, most of the successful ventures incorporated the elements (a) to (c) above, as revealed by the recent studies of 'success stories'. In these cases, both production and entitlement options have been enhanced without unduly damaging the resource base and environment (Jodha et al. 1992).

In the fragile, high-risk production environments, the essence of improved food security is the enhancement of the range and quality of production and exchange options, that help people to have command over multiple possibilities *vis-à-vis* food requirement. In the light of the above, option-enhancement strategies should be guided by the following considerations.

(a) In today's changed context, food security extends far beyond the stability of a self-provisioning system, that mainly focused on the production and consumption of food. Now, it should focus more on people's command over options reflected through what is described as entitlements. The strategy directed at increased entitlements would be different from that addressing the stability of subsistence supplies dominated by directly consumable items. Thus food security is directly linked to employment- and income-generation activities.

(b) Option enhancement (in the context of currently visible negative changes in the fragile resource areas) would require the following:

(i) Rehabilitation and upgrading of resource base and dependent production systems using nature's regenerative processes, usage regulation and application of modern science and technology.

(ii) Diversification of resource use, production activities (or rather income-generation activities) and demand itself. Again, in the changed context, diversification cannot be confined to biomass (including food) production activities but will have to incorporate interlinked activities of the primary, secondary and tertiary sectors, as experiences from most of the transformed areas have shown.

(iii) Intensification is an essential precondition for increased production from a resource base. To avoid inappropriate and resource-degrading intensification, the focus has to be on intensification of a total production system (involving diverse and interlinked activities) as against intensification of selected components (e.g. single-crop within an overall cropping/farming system) that generate imbalances and lead to disintegration of production systems. Science and technology has to play a crucial role in this.

(iv) Inter-system or inter-regional linkages is another step towards option enhancement. These steps (diversification, intensification, etc.) can help enhance production; but the gains from them (for increased entitlement) can be increased only with appropriate value-adding activities and exchange. This requires effective and equitable exchange relations and inter-system linkages based on comparative advantage. This is more so when harnessing of regional niche is involved.

(v) Entitlement enhancement through increased employment and income opportunities is the central purpose of development interventions. Nevertheless, because of the generalized nature of such interventions *vis-à-vis* diversities and location specificities characterizing fragile resource zones, such strategies did not succeed in the past. Hence, deliberate diversification of income- and employment-generating programmes and their adaptation to local circumstances is a key step towards ensuring food security (through entitlement enhancement).

(c) Demand management is an important aspect of food security. Micro-level demand management should cover elements such as diversity of food/food chain, recycling, sharing, consumption-production linkages, occupational diversification-based spatial and seasonal spread of demand pressure, etc. These unrecognized aspects of food security need strengthening. Admittedly, the formal food security arrangements cannot enforce such essentially informal measures. Just the same, formal interventions through technologies, decentralization, etc. can create objective circumstances to which people can respond and adopt such measures.

'Food deficit and food transfers' are important aspects of macro-level demand management in the fragile resource zones. Despite all efforts, food production may periodically fall short. Food transfer from surplus producing areas has been the government's response. This has two problems. First, in remote, less accessible areas such supplies are quite undependable. The focus on local production (with

Chart 4.7. Steps to Enhance Food Security in Fragile Zones

Focal Areas and Components	Implications and Action Areas
Growth in Physical Production	
Resource rehabilitation	Diversification, regeneration, recycling, usage regulation; Resource-focused reorientation of agricultural R&D incroporating elements of traditional systems
Resource upgrading	Conservation, stabilization, reforestation, water harvesting, irrigation measures; public investment, people's participation for small-scale, diversified activities; policy reorientation as required
Restoring integrity of production systems	Diversified, interlinked land-based activities, complementarity of: intensive/extensive land uses, annual-perennials; reorientation of agricultural R&D, support to traditional measures, focus on resource regeneration, high productivity, biomass stability
Higher productivity	Intensification of production systems, i.e. all interrelated activities rather than individual components; mixing short-duration annual-perennials; reorientation of agricultural R&D based on understanding of intensification needs; appropriate support systems
Entitlement Enhancement	
Going beyond biomass-centred, self-provisioning	Harnessing niche, high-value products, agro-business orientation, support systems and policy orientation to suit local diversity
Gains from inter-system links	Inter-regional exchange based on comparative advantages and complementarities, equity of exchange, itegrating primary, secondary tertiary sector activities; infrastructural support and policy reorientation
Fair share in harnessing of niche	Equitable product pricing, fair compensation for resource use, required policy change and local participation
Percolation of developmental gains	Reorientation of development interventions, location specific focus; sensitivity to local needs, decentralization, local participation

Note: For further details see Jodha 1991a,b, 1990a; Jodha and Partap 1993; Banskota 1989.

traditional features such as diversification, recycling, etc.) is still the best strategy there. In more accessible areas, agro-business-oriented activities based on niche can prove more effective for ensuring food security through 'entitlement enhancement'. The second aspect of food transfer is that due to their frequent foodgrain deficit, the fragile resource zones are often treated as liabilities for the national economy. This distorts the policy-makers' perception *vis-à-vis* food security in fragile resource zones, imparting it a charity orientation (Gadgil et al. 1988; Banskota and Jodha 1992b). However, if facts such as the off-site impact (as a cost) of decline of fragile resource zones due to inappropriate land use for increased foodgrains and the net transfer of resources from these areas to the mainstream economy (due to under-pricing, inadequate compensation, etc.) are considered, the 'liability'-dominated perspective will change (Anderson and Jodha 1994).

Chart 4.7 summarizes the practical implications of the issues relating to food security discussed above. The possibilities of food security through increased production and/or entitlement enhancement, using these steps, are corroborated by the experiences of agricultural and rural transformation in mountain areas such as Himachal Pradesh (India), Miyi and Ningnan counties (West Sichuan, China), Ilam district (Nepal), as shown by ICIMOD studies. Finally, however, it should be admitted that owing to the high degree of fragility and the limited carrying capacity of fragile resource zones, despite their conservation, upgrading and regenerative production practices, these regions may not be able to accommodate the rising number of people (and animals) for ever. Hence, control over internal and external pressure on resources is essential.

II. CPR Perspective

5

Common Property Resources in Crisis*

 Mixed success of even well-supported rural development interventions has been a long-standing concern of development agencies, including donor agencies. The failures are more conspicuous where the central role in the development processes was played by institutional as against technological factors; group action as against individual initiatives; participatory processes as against externally controlled and supported activities; management and sustainable use of resource base as against short-term productivity promotion; fragile and high-risk environments as against stability-wise prime areas. Responses to such failures often repeat history in one or another way, through (a) emphasizing technological remedies to handle primarily institutional problems; (b) strengthening physical infrastructure and imposing formal legalistic arrangements as substitutes for people's organizations; (c) bribing people through a variety of subsidies to induce their participation; and (d) (in some cases) dropping the problem altogether by declaring it a situation beyond redemption (using the inferences of 'feasibility' exercises).

Without belittling the importance of technology, support systems and incentives, it may be stated that the misplaced emphasis on such responses distorts the perspectives on the problems and possible solutions. Most importantly, they tend to block the opportunity to assess properly the ecological and institutional environment of the situation and understand people's traditional resource management and production strategies, which in turn can offer useful leads for evolving

*First presented as 'Common Property Resources: Contribution and Crisis', Foundation Day Lecture, 16 May 1990, Society for Promotion of Wasteland Development (SPWD), New Delhi. Reprinted in the *Economic and Political Weekly* (Quarterly Review of Agriculture) 25(26), (1990). A revised version was published as World Bank Discussion Paper no. 169, 1992.

workable options (in the situations considered 'hopeless') by the conventional development approaches.

The disregard of common property resources (CPRs) in rural development programmes in most developing countries is illustrative of the conventional approach. In the dry tropical regions of India, not only have public policies and programmes failed to recognize and harness the potential of CPRs, but have in effect operated against them.

Common property resources (CPRs) may broadly be defined as those (non-exclusive) resources in which a group of people have co-equal use rights. Membership in the group of co-owners is typically conferred by membership in some other group, generally a group whose central purpose is not the use or administration of resource (*per se*), such as village, tribe, etc. (Ostrom 1988, Bromley and Cernea 1989).

CPRs are an important component of natural resource endowment of rural communities in developing countries, making a significant contribution to the sustenance of rural households and offering a potential role in equitable and participatory development processes. Yet CPRs have generally been ignored by researchers, policy-makers, development planners and donors alike. Even when development interventions are focused on a community's physical/natural resource (e.g. village pasture, forest, etc.), their CPR dimension is seldom considered. Consequently, despite technological input, financial support and legal and administrative back-up, CPRs continue to degrade and disappear. The tragedy of commons is used as a convenient cover for the default and insensitivity of public interventions to these resources (Feeny et al. 1990). The present discussion highlights some of these issues and their policy implication for India's arid and semi-arid tropical parts. These regions, extending over more than 150 districts, account for above 43 per cent of India's total geographic area and 31 per cent of its total rural population (Jodha 1989a).

CPRs in Indian villages include community pastures, community forests, wastelands, common dumping and threshing grounds, watershed drainages, village ponds, rivers and rivulets with their banks and beds. Community pastures, community forests and wastelands, being large in area and major contributors to rural people's sustenance, are more important. Even where their legal ownership rests with some other agency (e.g. wastelands belong to revenue department of the state), *de facto* CPRs belong to the village community.

In developing countries, CPRs continue to be a significant

component of the land resource base of rural communities, more so in the relatively high-risk, low-productivity areas such as the arid and semi-arid tropical regions of India and several African countries. Historically, among the circumstances that favoured the institution of CPRs in these areas were: the presence of factors less favourable to rapid privatization of land resources, community-level concerns for collective sustenance and ecological fragility, and dependence of private-resource-based farming on the collective-risk-sharing arrangements (see Chart 5.1). CPRs in turn contribute to the production and consumption needs of rural communities in several ways and help satisfy several environmental imperatives of dry-land conditions. Nevertheless, there is a crisis of CPRs, reflected in their area shrinkage, productivity decline and management collapse. Corroborating the field-level evidence presented here are other micro-level studies, such as Iyengar 1988, Brara 1987, Chen 1988, Blaikie et al. 1985, Gupta 1986, Wade 1988, Ananthram and Kalla 1988 and Oza 1989.

The following section introduces the study areas, data base and methodological highlights of the field work behind the evidence presented here. Section 3 presents quantified details on the contributions of the CPRs. Section 4 discusses the changing status of CPRs in terms of decline in area, productivity and management systems. The dynamics of decline of CPRs discussed in Section 5 identifies the role of various factors in this process. Section 6 discusses the people's responses to the changing situation of CPRs. Section 7 comments on the prospects of CPRs in dry regions of India. Section 8 discusses the policy implications of major issues highlighted by the essay, with focus on advocacy and action to rehabilitate CPRs.

DATA BASE AND METHODOLOGY

AREA AND INFORMATION COVERAGE

The field studies, conducted during 1982–6, covered 82 villages from 21 districts, scattered in seven major states in the dry tropical zone of India (see Map 5.1). Annexure 5.1 provides some agro-ecological, demographic, administrative and CPR-related details of the study areas.

The method of collecting information depended upon the nature of information. The methods included regular monitoring, structured surveys, physical verification/measurement, recording of oral history and participant observations by (background and age-wise

Chart 5.1. Circumstances Historically Associated with the Importance of CPRs[a]

Natural Resource Base and Agro-Ecological Features

Low and variable precipitation, heterogeneous (including sub-marginal fragile land) resources, nature's low regeneration capacities, limited and high-risk production options, etc.

Implications/Imperatives

Regional Level	Community Level	Farm Level
Low population pressure; market isolation; limited technological, institutional interventions, etc.	Heterogeneity, fragility of resource base; inadequacy of private risk strategies;	Narrow, unstable production base; diversified, biomass centres, land-extensive farming systems
Limited incentives and compulsions for privatization of CPRs	Balancing extensive-intensive land uses; focus on collective risk sharing	Reliance on collective measures against seasonality and risk
Circumstances favourable to CPRs	Community sanctions for CPRs (protection, access, usage, etc.)	Complementarity of CPR-PPR[b] based activities

[a]For more details and evidence on different aspects covered by the chart see Chapter 2, Jodha 1988b, 1991a, M. Gadgil 1985, Gupta 1986, Chambers et al. 1989b, Gadgil and Iyer 1989, Arnold and Stewart 1991.
[b]PPR = Private Property Resources.

Map 5.1. Districts and Number of Villages Covered by the Study on CPRs

heterogeneous) teams of formal and informal cooperators in each district. The information thus collected was supplemented with detailed longitudinal data available from ICRISAT's village-level studies (Walker and Ryan 1990) conducted in ten villages of five districts, which were also covered by the CPR studies. The number of units (e.g. villages, households, CPR-units, etc.) covered by different investigations will be indicated while reporting the results.

The study areas were selected purposively, with two preconditions: (a) representation to the zones with different soils, agro-climatic features and population densities; and (b) availability of local cooperators to help in the field work, the latter being a logistic requirement imposed by the nature of the studies. The CPR studies, unlike routine agro-economic surveys, required greater flexibility and use of unconventional methods of information gathering on the one hand and close familiarity of investigators with the villages and their oral history on the other. Because of the latter, identification of relevant cooperators from different agro-climatic zones preceded the purposive selection of study areas. Apart from the above features, the approach and methodology of CPR research was unconventional in several other ways. The important ones may be stated below.

THEORETICAL FRAMEWORK

While acquaintance with received theories and concepts in the fields of property rights and regimes was acquired through literature on the subject, the dictates of the theories were not allowed to constrain the approach of studies. Accordingly, rather than trying to confirm or reject specific hypotheses, the focus of the research was on past and present status, management and productivity of community resources in the villages. Even the legal provisions that bestowed ownership of a given communal resource to any agency (e.g. village wasteland with the state's department of land revenue) were ignored by treating such resources as part of CPRs, because *de facto* they qualified for such treatment. This approach provided flexibility to cover issues as they emerged with the progress of field work for nearly four years. Consequently, the enquiries initially focused on the role of CPRs in private-resource-based farming systems, and finally covered vast and varied areas including CPRs *vis-à-vis* issues of environment and ecology; poverty and equity; institutional and technological interventions; rural factionalism and participatory development; grassroots democracy and bureaucratic perceptions, and finally the whole dynamics of rural transformation.

QUESTION OF UNDERSTANDING AND KNOWLEDGE

The focus of the work was first on understanding different aspects of CPRs, especially as the villagers would look at them. This was followed by documentation and collection of information. This meant successive application of research methods ranging from rapid rural appraisal (RRA), or rather participatory rural appraisal (PRA), to structural survey for recording the facts in quantitative terms. This was facilitated by collaboration and involvement of people with different background and skill. Most of the village-level cooperators in the study had not only witnessed the changes in CPR situation but also participated in the process of change and faced its consequences.

MIX OF METHODS FROM ANTHROPOLOGY AND ECONOMICS

Another key feature of the methodology of field investigations, as alluded to earlier, was the combined use of research methods and approaches usually associated with different social science disciplines. In particular, approaches followed by anthropologists and rural socio-logists were emphasized for three reasons. First, it helped in better understanding of oral history and people's perceptions *vis-à-vis* CPRs. Second, due to the primacy of institutional dimensions in CPR issues, anthropological approaches proved more helpful. Third, it helped in the integration of the available records and information with the socio-cultural-economic processes at micro and macro levels. The above approaches were supplemented by methods used for agro-economic investigation, involving recording of quantitative dimen-sions of CPR issues. Depending on the nature of issues covered (e.g. change in the area of CPRs or seasonal collection of fuel/fodder from CPR plots) and availability of cooperators with requisite skills, quanti-tative data were collected with different frequencies (e.g. ranging from once a year to once a week). Complementary use of different social science methods (supplemented by a few simple agronomic methods) paid well in terms of better understanding, documentation, analysis and communication of CPR situation in the study areas. The awareness and impact these studies have so far generated in policy and programme contexts, should largely be attributed to this fact (SPWD 1991, Arnold and Stewart 1991, GOI 1990). However, the key limitation of the approach is that it is too demanding in terms of workers' personal commitment, their shared concerns, and of course time and patience. Field research conducted in 'project mode' including that supported by the donors, which have rigid TOR (terms of reference), etc., often does not have design and mechanisms to ensure the above prerequisites.

FLEXIBLE LOGISTICS

Most of the attributes of the CPR research project stated above, including the adoption of unconventional methods, flexibility, etc., became possible because the researchers were not strictly bound by any TOR except fulfilment of some basic conditions, such as confinement of the studies to the dry regions of India, covering major soil-rainfall zones within the dry regions, and identification of complementarities between CPR- and PPR-based farming systems.

A small grant from the Ford Foundation that supported this study also did not have rigidities usually associated with research grants. ICRISAT's logistic support was available but its use was reduced to the minimum by depending on district- and village-level local facilities. This was in keeping with the useful tradition of anthropological investigations, where the absence of a gap between researcher and the respondents is a key element. Moreover, CPR studies involved a number of sensitive issues which could not have been covered by typical structured surveys.

CONTRIBUTIONS OF CPRs

In the dry region of India, the contribution of CPRs to village people's employment, income generation, and asset accumulation (directly or through complementing the private-resource-based activities) are numerous, but seldom recognized. This invisibility is even more pronounced in long-term social and environmental processes characterizing dry areas. Chart 5.2 gives a broad picture of the contributions of various CPRs.

QUANTIFICATION OF BENEFITS

Quantification of the contributions of CPRs is not easy, owing to a number of problems of monitoring and measurement. However, it was attempted in the field studies and the relevant information is summarized in Table 5.1.

Owing to their degradation and reduced productivity, CPRs do

Chart 5.2. Contributions of CPRs to Village Economy

Contribution	Community Forest	Pasture/ Wasteland	Pond/ Tank	River/ Rivulet	Watershed Drainage/ River Banks	River/ Tank Beds
Physical Products						
Food/Fibre items	✓	✓	✓	✓	✓	✓
Fodder/Fuel timber, etc.	✓	✓				
Water	✓					
Manure/Silt/Space	✓	✓	✓	✓	✓	
Income/Employment Gains						
Off-season activities	✓	✓				
Drought period sustenance	✓	✓	✓	✓		
Additional crop activities	✓	✓				
Additional animals	✓	✓				
Petty trading/handicrafts	✓	✓	✓			
Larger Social, Ecological Gains						
Resource conservation	✓	✓				
Drainage/Recharge of ground water						
Sustenance of poor	✓	✓	✓			
Sustainability of farming systems	✓	✓	✓	✓		
Sustainability of resource supply	✓	✓		✓	✓	
Better micro-climatic environment	✓	✓	✓	✓		

Source: Adapted from Jodha 1985a.

Table 5.1. Extent of People's Dependence on Common Property Resources (CPRs) in Dry Regions of India[a]

State (with No. of Districts and Villages)	Household Category[b]	CPRs Contribution to Household Supplies, Employment Income, etc.					Value of Gini-coefficient of Incomes from[h]	
		Fuel Supplies[c]	Animal Grazing[d]	Per household		CPR Income as Proportion[g] (%)	All Sources	All Sources Excluding CPRs (%)
				Employment Days[a] (No.)	Annual Income[f] (Rs.)			
Andhra Pradesh (1, 2)	Poor	84	-	139	534	17	0.41	0.50
	Others	13	-	35	62	1	0.41	0.50
Gujarat (2, 4)	Poor	66	82	196	774	18	0.33	0.45
	Others	8	14	80	185	1	0.33	0.45
Karnataka (1, 2)	Poor	-	83	185	649	20		
	Others	-	29	34	170	3		
Madhya Pradesh (2, 4)	Poor	74	79	183	733	22	0.34	0.44
	Others	32	34	52	386	2	0.34	0.44
Maharashtra (3, 6)	Poor	75	69	128	557	14	0.40	0.48
	Others	12	27	43	177	1	0.40	0.48
Rajasthan (2, 4)	Poor	71	84	165	770	23		
	Others	23	38	61	413	2		
Tamil Nadu (1, 2)	Poor	-	-	137	738	23		
	Others	-	-	31	164	2		

[a]The number of sample households from each village varied from 20 to 36 in different districts. 'Poor' are defined to include agricultural labourers and small farm (<2 ha. dryland equivalent) households. 'Others' include large farm households only.
[b]Fuel gathered from CPRs as proportion of total fuel used during three seasons covering the whole year.

^cAnimal unit grazing days on CPRs as proportion of total animal unit grazing days.

^dTotal employment through CPR product collection.

^eIncome mainly through CPR product collection. The estimation procedure underestimated the actual income derived from CPRs (Jodha 1986a).

^fCPR income as per cent of income from all other sources.

^gHigher value of Gini coefficient indicates higher degree of income inequalities. Calculations are based on income data for 1983–4 from a panel of households covered under ICRISAT's village-level studies (Walker and Ryan 1990). The panel of 40 households from each village included 10 households from each category, namely large, medium and small farm households and labour households.

not at present offer high returns. The rural poor, with limited alternative means of income, depend more on the low pay-off options offered by CPRs, while the rural rich (indicated by the category 'others' in Table 5.1), depend very little on CPRs. The proportion of poor households depending on fuel, fodder and food items from CPRs ranged between 84 and 100 per cent in different villages. The corresponding figures for rich households, except in very dry villages of Rajasthan, ranged between 10 and 19 per cent (Jodha 1986a). The recent tendency of the rural rich is to acquire CPR land as private property rather than to rely on meagre output from CPRs.

Table 5.1 suggests the following inferences. The rural poor receive the bulk of their fuel supplies and fodder from CPRs. CPR product collection is an important source of employment and income, especially in times of lean employment. Furthermore, CPR income accounts for a conservative estimate of 14 to 23 per cent of household income. More importantly, CPR income helps to reduce the extent of rural income inequalities, as indicated by lower values of the Gini-coefficient.

For the rural poor, per worker employment days generated by CPR-based activities were higher than the days of employment on their own farm or on public works under the anti-poverty programmes. The potential for self-employment and income generation can be enhanced through development and proper management of CPRs, as indicated by villages with better upkeep of CPRs compared to the others (Jodha 1986a).

Contributions to Private Farming

Table 5.2 highlights the complementary role of CPRs in the PPR (private property resource)-based farming systems. It is seen from the table that 31 to 42 per cent of the total own farm inputs used during the pre-sowing to pre-harvest stages of cropping are contributed by cash or kind inflows from CPRs. Such contributions during other stages of the cropping season are smaller because alternative means are available, such as high wage earnings, etc. A still greater dependence of private resource-based crop farming on CPRs is revealed by the extent of support it receives for the sustenance of farm animals. Maintaining these animals without CPR would have implied diversion of a substantial proportion (48–55 per cent) of crop lands from food and cash crops to fodder crops. The alternative

Table 5.2. CPRs' Contribution to PPR-based Farming Systems

Item	Range of Values in Different Areas (%)
A. Cash/kind inflows in total own inputs used during different stages of cropping[a]	
Pre-sowing to pre-harvest	31–42
Harvest	11–16
Post-harvest	8–10
B. Potential decline in own resource availability for cropping in the absence of CPRs[b]	
Draft power	68–76
Manure/dung	35–43
Land area for cash/food crops	48–55
Crop by product for sale	84–96
C. CPR contribution to total sustenance income (excluding relief and credit) during:[c]	
Drought years	42–57 (68–72)[d]
Non-drought years	14–22 (25–38)

[a]Data under sections A and B covering average of two cropping years (1983 and 1984) relate to small and marginal farmers (i.e. those having <2 ha. dry land equivalent of area). The districts and number of sample households covered are as follows: Mahabubnagar (13), Akola (10), Sholapur (12), Sabarkantha (20), Raisen (18).

[b]Procedure for estimating potential decline in own resource supplies (following the non-availability of CPRs) was as follows. (i) Average fodder requirement and output of small number of animals currently stall fed for 6 to 8 months a year were estimated. (ii) This average was applied to currently owned animals receiving negligible stall feeding, to estimate their fodder requirement and its implications in terms of transfer of own land area from cash and food crops to fodder crops and reduced marketable surplus of crop by-products. In the absence of above potential adjustments, the implications in terms of reduced animal numbers and consequent decline in draft power and manure supplies were estimated.

[c]Data based on studies of drought years and post-drought years conducted in the following districts (with number of sample households), Banaskantha in Gujarat (100), Barmer and Jodhpur, in Rajasthan (144 and 100 respectively), Sholapur in Maharashtra (80). For details see Jodha (1978).

[d]Figures in the parentheses indicate percentage of village households using CPRs.

Source: Adapted from Jodha 1990a.

option, i.e. reducing the number of animals to a level sustainable by own fodder/feed resources, would have implied loss of own farm inputs, e.g. draft power (68–76 per cent) and farm yard manure (35–43 per cent).

Table 5.2 also indicates the CPRs' contribution to drought period sustenance of the farming households. If the help received through relief and credit is excluded, 42–57 per cent of the total sustenance income during the drought years is contributed by cash and kind inflows from CPRs. The corresponding figures for non-drought years (post-drought years) are 14–22 per cent. The key inferences relating to CPR–PPR complementarity as seen from Table 5.2 are as follows.

(i) Due to short wet period (planting period) and quantity of manure required for his land, the dry-land farmer keeps more animals than could be maintained or fully utilized over the year, by his narrow production base consisting of small holding and short cropping season. The implied high overhead cost of private crop farming is met through CPRs as a source of fodder and forage.

(ii) Owing to non-covariability of production flows (and input requirements) of CPR-use and PPR-based farming, CPRs help fill in the resource and product gaps faced by private-resource based farming.

(iii) The pressure on CPRs is greater when the productivity of PPR-based farming (as during the drought years) is low. This is also true in the spatial context when areas with high- and low-cropping potential are compared.

(iv) PPR-based dry-land farming can be strengthened through the revitalization of CPRs.

COLLECTIVE AND ENVIRONMENTAL GAINS

A number of benefits of CPRs to the whole village rather than to individual households were also recorded for some villages where selected CPRs were still managed well. Such villages had self-sufficiency and surplus production of off-season vegetables and hence better nutrition (due to better management of river/rivulet beds/banks); had lesser incidence of run-off or drought-induced crop-resowing and crop failure (due to better management of watershed through natural vegetation, tamed drainage and soil working); had no drinking-water problem (due to collective upkeep of watering points); had dependable water supply from dug wells even during the drought years (due to better soil works and management of watershed and fields in the catchment); had high-income-generating cottage industries based on CPR products (gum, wild fruits, fibre, etc.); had income (in cash or kind) from CPRs to maintain the village bulls and pay for improved facilities in village school; and had lesser dependence on government grants and relief. These and a few other less quantifiable gains, such as those relating to the environmental situation (e.g. protection of young trees, unseasonal lopping of trees, etc.) came to light while preparing the village profiles during the field studies. It should be noted, however, that all the above features do not relate to the same set of villages. Instead, there were 33 out of a total of 82 villages covered by the study, which had one or more of the above features.

CPR GAINS UNPERCEIVED

Despite visible and non-visible, short-term and long-term, individual and collective gains from CPRs, these resources remained a most neglected area in development planning. Whether in employment- and income-generation programmes, applied nutrition programmes, poverty eradication and equity promotion programmes, resource management and environmental stability programmes for the dry regions, no Plan documents make explicit reference to CPR contributions and approach to harness them. Public intervention treats a village community's natural resources merely as biophysical entities, with little concern for their CPR dimension and its implications. Finally, owing to the policy-makers' and planners' ignorance or deliberate disregard of CPRs, a number of welfare and development interventions have been initiated that have severe negative side-effects on CPRs (to be discussed later). A major dimension of rural realities is thus being missed in development strategies for dry areas.

DEPLETION OF CPRs

Disregard of CPRs and their contribution both by welfare and production programmes, has not only led to their marginalization as a useful resource but has caused their depletion both in terms of area and productivity. This in turn induces a further fall in their pay-off, to be followed by further neglect and degradation. Decline in area of CPRs is relatively easy to observe with the help of written or oral records of village land usage. In contrast, the fall in production of CPRs, although keenly felt by villagers, is difficult to quantify because their productivity has not been recorded in the past.

DECLINE OF CPR AREA

Changes in area have been recorded for community pastures, village forests, wastelands and minor CPRs such as community threshing grounds, watershed drainages and fallowed catchments of ponds in all the 82 villages covered by the study. The reference period is 1950–2 when comprehensive land reforms were introduced in the country. The introduction of land reforms initiated changes in the status and management pattern of CPRs. Moreover, being a major event in the memory of villagers, land reforms also provided an important context that facilitated recording of oral history of CPRs. The 1950–2 situation of CPRs is compared with that during 1982–4, when the field work was conducted. According to Table 5.3, CPR area has declined by 31 to 55 per cent in the study villages. The impact of this change is clearly visible in terms of both decline in proportions of CPR lands in total village area and increase in population pressure on CPRs. Substantial decline in the extent of CPRs has also been recorded by several other studies in different parts of the country (Iyengar 1988; Blaikie et al. 1985; Brara 1987; K. Chopra et al. 1990).

At the village level the process of privatization of CPR lands had all the following elements. It involved gradual extension of private field borders into adjoining CPR land; outright grabbing of CPR plots by village influentials individually or collectively, etc. Such cases are legalized by the government in due course. Another form of privatization was distribution of CPRs land as private land by the government (more on this later). Yet another way in which CPR area was curtailed involved acquisition of such lands by the government for development and resource harnessing through its own agencies, e.g. forest department or contractors, or for using such lands for other public facilities such as the village panchayat complex, statues and local parks in the name of departed leaders.

PHYSICAL DEGRADATION OF CPRS

In the absence of recorded benchmark information for assessing degradation or decline in productivity of CPRs over time, a benchmark had to be constructed from oral history and scattered village records. Evidence on reduced productivity and production potential of CPRs is summarized under Table 5.4.

A drastic decline in the number of products, following the disappearance of a number of plant and tree species from CPR

Table 5.3. Extent and Decline of CPR Area

State (and no. of districts)	No. of Study Villages	Area of CPRs 1982–4 (ha)	CPRs Area as % of Total Village Area		Decline in Area of CPRs since 1950–2 (%)	Persons per 10 ha. of CPR area (no.)	
			1982–4	1950–2		1951	1981
Andhra Pradesh (3)	10	827	11	18	42	48	134
Gujarat (3)	15	589	11	19	44	822	238
Karnataka (4)	12	1165	12	20	40	46	117
Madhya Pradesh (3)	14	1435	24	41	41	14	47
Maharashtra (3)	13	918	15	22	31	40	88
Rajasthan (3)	11	1849	16	36	55	13	50
Tamil Nadu (2)	7	412	10	21	50	101	286

Note: CPRs include community pasture, village forest, wasteland, watershed drainage, river and rivulet banks and other common lands. Data indicate average area per village.

Source: Adapted from Jodha 1986a, where more disaggregated details are reported.

Table 5.4. Some Indicators of Physical Degradation of CPRs[a]

Indicators of Changed Status and Context for Comparison		State (with no. of villages)							
		Andhra Pradesh (3)	Gujarat (4)	Karnataka (2)	Madhya Pradesh (4)	Maharashtra (3)	Rajasthan (4)	Tamil Nadu (2)
CPR products collected by villagers (No.):[b]	In the past	32	35	40	46	30	27	29
	At present	9	11	19	22	10	13	8
Per hectare no. of trees and shrubs in:	Protected CPRs[c]	476	684	662	882	454	517	398
	Unprotected CPRs	195	103	202	215	77	96	83
No. of watering points (ponds) in grazing CPRs	In the past	17	29	20	16	9	48	14
	At present	4	13	4	3	4	11	3
No. of CPR plots where rich vegetation, indicated by their nomenclature, is no more available	-	12	3	6	4	15	-	
CPR area used for cattle grazing in the past, currently grazed mainly by sheep/goat (ha)[d]		48	112	95	-	52	175	64

[a] Based on observation and physical verification of current status (during 1982–4) and the past details collected from oral and recorded description of CPRs in different villages. The choice of CPRs where plot based data are reported was guided by availability of past information about them.

[b] Includes different types of fruits, flowers, leaves, roots, timber, fuel, fodder, etc. in the villages. 'Past' indicates the period preceding the 1950s 'Present indicates the early 1980s.

[c] Protected CPRs were the areas (called 'oran', etc.), where for religious reasons live trees and shrubs are not cut. The situation

of CPR plots (numbering between 2 to 4 in different areas) was compared with other bordering plots of CPRs which were not protected by any religious or other sanctions.

[d]Relates to area covered by specific plots, traditionally used for grazing high productivity animals (e.g. cattle in milk, working bullocks or horses of feudal landlords). Because of their depletion, such animals are no more grazed there.

Source: Adapted from Jodha 1990a.

lands, which villagers used to gather from the commons in the past (i.e. before the early 1950s), is a major indicator of physical degradation of CPRs. Local names of a limited number of plots indicated their coverage by specific vegetation in the past. Hardly any of those species grow there now. Similarly, a number of selected CPR plots traditionally used for grazing more productive animals, like lactating cattle, working bullocks, etc., are no more able to carry these animals. With their forage potential depleted and vegetative composition changed, sheep and goats graze them instead of cattle. The number of watering points, an important component of common grazing lands, has also declined.

The difference in the number of species found on protected and unprotected CPRs is an important indicator of vegetative degradation and associated resource depletion. Certain CPR areas are protected through religious sanctions against removal of live trees, shrubs, etc. The per hectare number of trees and shrubs recorded was three to six times higher in protected CPRs compared to unprotected ones (Table 5.4).

An important indicator of reduced productivity of CPRs is the greater time and longer distances involved in collecting the same or less quantity of CPR products today than in the past. Again, in the past the whole village community used CPRs; now, as indicated earlier, mainly poor households, with few alternative options, try to meet their needs from meagre products from these resources.

Reduced production potential of CPRs has also been recorded by already mentioned other studies. The negative changes in the resource base are indicators of the worsening village environment due to the decline of CPRs.

SLACKENING OF MANAGEMENT SYSTEMS

CPRs get physically degraded on account of over-exploitation and poor upkeep. Both reduction in their area, leading to overcrowding, and absence of usage regulation have encouraged over-exploitation of CPRs. The inability to enforce obligations of CPR users (in terms of grazing tax or compulsory labour input for trenching, fencing, etc.) has led to their poor upkeep. These failures have resulted from slackening or abolition of the traditional formal/informal management practices for CPRs (Jodha 1985b; Singh 1986; Arnold and Stewart 1991). Table 5.5 presents the extent of discontinuation of several measures which constituted integral parts of traditional management

Table 5.5. Some Indications of Changes in Management of CPRs: Number of Villages Pursuing Given Measures

State (with no. of villages)	Formal/Informal regulations on CPR use[a]		Formal/informal taxes/ levies on CPR use[c]		Users' formal/informal obligation towards upkeep of CPRs[d]	
	In the past[b]	At present[b]	In the past	At present	In the past	At present
Andhra Pradesh (10)	10	-[c]	7	-	8	-
Gujarat (15)	15	2	8	-	11	2
Karnataka (12)	12	2	9	-	12	3
Madhya Pradesh (14)	14	2	10	-	14	3
Maharashtra (13)	11	1	6	-	10	1
Rajasthan (11)	11	1	11	-	11	2
Tamil Nadu (7)	7	-	4	-	7	1

[a]Measures such as regulated/rotational grazing, seasonal restrictions on use of CPRs, provision of CPR watchmen, etc.

[b]'Past' stands for period prior to the 1950s, present stands for the early 1980s.

[c]Measures such as grazing taxes, levies and penalties for violation of regulations on use of CPRs. See Jodha (1985b) for descriptive account.

[d]Measures such as contribution towards desilting of watering points, fencing, trenching, protection of CPRs, etc.

[c](-) indicates nil.

Source: Adapted from Jodha 1990a.

systems of CPRs. In most states, not even a tenth of the villages have kept up their practices conducive to CPR management.

CPRs AND THE PROCESS OF PAUPERIZATION

In the dynamic context, depletion of CPRs and implied decline in access to bio-mass is an important indicator of rural pauperization. One aspect of this pauperization is that the functions and contributions of CPRs, implied by Tables 5.1 and 5.2, have declined or become less feasible. The pauperization process however, involves more than this.

First, in the larger social and ecological contexts, the transfer of sub-marginal CPR lands to crop cultivation, through their privatization, implies a step towards long-term non-sustainability of land-based activities in dry regions (Jodha 1986b, 1989b).

Secondly, reduced products and income-generation options, following degradation of CPRs, imply increased scarcity, and stress for those who depend on CPRs. The longer time and distance involved in collecting the same or lesser quantities of CPR products and reduced effective period (months) of sustained grazing offered by CPRs today, as compared to the past, are just two of the several examples of this phenomenon.

Thirdly, despite the increasingly inferior options available from CPRs, the rural poor continue to depend upon them because the opportunity cost of their labour to harness the inferior options has become even lower. The progressive decline in the value of CPR products, accompanied by the increasing number of people relying on them for sustenance, is a clear indicator of increasing poverty. In this process, where the community silently eats away its permanent asset, the aftermath may prove costlier than if the public exchequer had found some other ways of helping the poor. The cost of the process is invisible because the poor are sustained by the CPRs without any direct and visible burden on the public exchequer, such as through community subsidy or development assistance.

CPR–POVERTY–ENVIRONMENT LINKS

The final consequence of the vicious cycle involving poverty and CPR degradation is the disruption or elimination of vital biophysical processes (e.g. regeneration, nutrient and moisture flows, etc.), which sustain the physical productivity of natural resources in dry regions (Jodha 1991a). The state's efforts to control this decline, to be discussed in the next section, have not succeeded because they lacked CPR perspective.

DYNAMICS OF THE DECLINE OF CPRs

The decline in area, productivity and upkeep of CPRs is common to most developing countries, where these resources continue to be important. Recent literature attributes these changes to population growth, market forces, public intervention, technological changes and environmental stress (e.g. drought) (see Runge 1981; Repetto and Holmes 1984; Bromley and Chapagain 1984; Jodha 1985a, b; Bromley and Cernea 1989). Chart 5.3 sketches the process through which these factors individually or jointly contribute to the decline and depletion of CPRs. These factors influence the informal or formal norms and arrangements governing people's approach to CPRs. These norms and arrangements can alter with changes in the perceptions and needs of the community. These changes in turn are reflected through public policies and interventions and local communities' responses to them. This is being focused in the following discussion. Ideally, change in CPRs would be discussed in terms of the process sketched in Chart 5.3, but detailed data not being available for each step of the process, we shall concentrate largely on public intervention affecting CPRs.

PUBLIC INTERVENTION AFFECTING CPRs

Public policies and programmes influencing CPRs may be categorized as: (a) those affecting the area of CPRs; (b) those relating to products and productivity of CPRs; and (c) those influencing the management, usage and upkeep of CPRs. The categories are not watertight compartments.

PUBLIC INTERVENTION AND DECLINE OF CPR AREA

It is seen from Table 5.5 that in all the areas studied large-scale privatization of CPRs has led to decline in their extent. This change goes in tandem with the government's land distribution policies, which introduced land reform in the early 1950s and followed it up to date with various populist programmes. Having failed to acquire land for redistribution through land ceiling laws (Ladejinsky 1972) or through voluntary donation (by private landowners), under

Chart 5.3. Process of Depletion of Common Property Resources in the Dry Regions of India

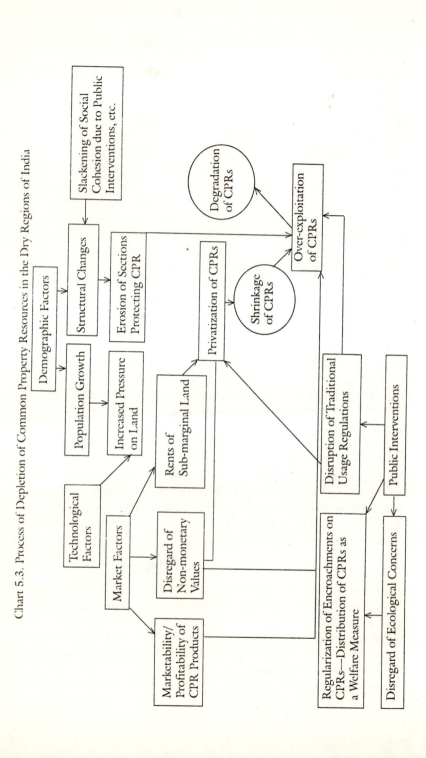

movements like *Bhoodan*, curtailment of CPR lands was seen as the easiest option out. Privatization of CPR lands involved either (a) formal distribution of land to landless and other groups under different welfare and development schemes, or (b) legalization of illegally grabbed CPR lands by people.

In the garb of helping the poor, however, privatization of CPRs brought more land to the already better-off households. It may be seen from Table 5.6 that whereas the proportion of poor households among the total recipients of CPR lands was generally higher in all the study villages, the proportion of land they received was much lower (barely one hectare per household) than other social groups (between two and three hectares). Those who already had relatively more land also received more out of the privatized CPR land (see col. 9). Furthermore, detailed enquiry revealed that poor households in all but one area were dispossessed of 23 to 45 per cent of the land received. The reasons included lack of complementary resources to develop and use the land, quality of land too poor to sustain annual cropping, etc. (Jodha 1986a). Thus the government's well-meant policies of helping the rural poor through land distribution were misconceived. It is doubtful whether the poor people's collective loss through reduced CPRs has been compensated by their individual ownership of former-CPR lands.

Public Intervention and CPR Productivity

The policies and programmes for raising the productivity of CPRs, adopted since the early 1950s, generally lacked CPR perspective. For instance, even when the programme titles refer to the community (e.g. rehabilitation of community forest or community pasture), they are often treated as state-run activities. They are conceived as measures related to physical resources located in the villages, implemented by administrative-cum-legal procedures, and sustained by state subsidy, with hardly any involvement of the village community. Chambers et al. (1989b) summarize evidence on these aspects from different studies. Another related feature of the productivity-raising initiatives for CPRs is their almost exclusive focus on production technology. Having a strong input from relevant science and technology, these programmes emphasize techniques rather than community involvement and user perspective. Hence, one comes across the long inventories of technically well assessed species of trees and grasses, methods for reseeding rangelands and reforesting wastelands, plant

Table 5.6. Distribution of Privatized CPR Lands to Different Household Groups in Dry Regions of India[a]

State (with no. of villages) 1	Total Land Given (ha) 2	Total Households Receiving Land (no.) 3	Share of Poor[b] in (2) (%) 4	Proportion of Poor in (3) (%) 5	Per Household Land Received by[c]		Average Land Size after Receiving New Land	
					Poor (ha) 6	Others (ha) 7	Poor (ha) 8	Others (ha) 9
Andhra Pradesh (6)	493	401	50	74	1.0	2.1	1.6	5.0
Gujarat (8)	287	166	20	45	1.0	2.6	1.8	9.4
Karnataka (9)	362	203	43	65	1.3	3.0	2.2	8.0
Madhya Pradesh (10)	358	204	42	62	1.2	3.2	2.5	9.5
Maharashtra (8)	316	227	38	53	1.1	1.9	2.0	6.2
Rajasthan (7)	635	426	22	36	1.2	3.2	1.9	7.2
Tamil Nadu (7)	447	272	49	66	1.0	1.5	1.9	6.7

[a]Number of districts covered by the table are 3 in each of the states except Andhra Pradesh and Tamil Nadu where 2 districts each were covered.
[b]'Poor' includes agricultural labourers and small farm (<2 ha of dry land equivalent) households.
[c]'Others' in this table unlike other tables, includes both medium and large farm households.
Source: Adapted from Jodha 1986a.

establishment and thinning techniques, and a variety of other silvicultural or agrostological recommendations for community lands. There is little, however, in terms of institutional sensitivity in these measures. Even more ironically, to establish and demonstrate the viability of technological measures, community lands are often alienated from the people, and transferred to pilot projects, etc. (Chambers et al. 1989b).

Even donor-supported initiatives on raising the productivity of CPRs share these features. For instance as a part of the World Bank-supported project (during the early 1970s), called 'drought proofing of drought-prone areas', pasture development was focused on in the districts of Nagaur and Jodhpur (Rajasthan). The present author visited nearly a dozen CPR units (called *jod, beed*, etc.) in these districts at two points of time. The project provided for fencing, new plantations, soil works, etc., in these CPR units, and facilities of watchmen and panchayat budget for the purpose. The key decision-makers in the project were district-level officials of soil conservation, planning and statistics departments, and important figures of district and village panchayats and a few village influentials who managed to present their private land as CPRs to get the benefit of the project. Villagers in general were unhappy with the project. Not only had it reduced their usable resources but it did not also provide a forum for venting their complaints and views. When the project ended (during the second visit of the author), in eight out of thirteen cases, all the physical and institutional arrangements provided by the project had disappeared. Of the remaining five cases (with visible impact of the project), three belonged to the influential individuals as their private grazing-cum-fodder-collection fields (beeds). One might recall here the unsuccessful livestock and pasture development schemes in different parts of Africa (Bromley and Cernea 1989), and note the commonalities of failure between the programmes.

A serious consequence of productivity-raising efforts initiated without sufficient concern for user perspective is the virtual conversion of CPR lands into commercial production fields, as witnessed in a number of social forestry projects (Chambers et al. 1989b; Arnold and Stewart 1991; Gupta 1986). In the process, most of the functions of CPRs indicated by Table 5.2 are sacrificed. The more productive CPRs face yet another problem in that the state often attempts to grab them, directly or through contractors, acquiring monopoly of product collection or marketing (Jodha 1985a). The villages' protests in some cases end up in prolonged litigation (Kaul 1987). Alienated from their own resources, the villagers adopt desperate measures and look for opportunities to over-exploit such resources.

PUBLIC INTERVENTION AND MANAGEMENT OF CPRs

A side-effect of certain institutional reforms, such as the introduction of land reform and new village panchayat system (elected village councils) has been that the traditional management systems for CPRs (involving usage regulation, enforcement of user obligations, and investment for conservation and development), have virtually disappeared. Land reform led to the abolition of a number of levies and taxes on CPR users (Jodha 1985b). The new institution of elected village panchayat undermined the traditional informal authority of village elders and replaced the formal authority of feudal landlords in some areas. Panchayats, despite their legal powers, are generally unable to enforce any regulation about CPRs. Their dependence on community votes prompts them to avoid unpopular steps like enforcing CPR-user obligations, and their domination by village influentials with little interest in CPRs makes them ineffective in terms of CPR perspective (Jodha 1985b; Gupta 1986; Arnold and Stewart 1989). At the same time, panchayats rarely miss an opportunity of seeking government grants in the name of CPRs. The panchayats' default has thus converted CPRs into open access resources with the consequent tragedy of open access resource (Bromley and Cernea 1989). The exceptions are where village elders still have informal authority (Brara 1987). Furthermore, traditional conventions and informal social sanctions relating to the use and maintenance of CPRs have been replaced with unenforceable legal and administrative measures. This has marginalized the people's initiatives and alienated them from CPRs. It has also encouraged dependence on government grants or relief rather than mobilize local resources (as in the past) for the upkeep of CPRs (Jodha 1985b).

Village panchayats thus become a replica in miniature of the state and formal source of (unenforceable) authority. Despite being part of the village, in most cases they become less sensitive to village realities (including declining CPRs) and more responsive to signals and temptations from above. Ironically, they become the focal point of externally conceived development intervention. No wonder in such cases, often with the involvement of NGOs, people seek alternative organizations (e.g. user groups) to manage community resources.

SIDE-EFFECTS OF OTHER DEVELOPMENTS

Components of general development strategies for rural areas also may adversely affect CPRs. The government-subsidized process of tractorization, for example, has led to rapid conversion of sub-marginal CPR lands into crop lands (Jodha 1974, 1985b). The increased monetization and commercialization of rural areas, a part of the transformation process, has adversely affected people's attitude to CPRs. Improved accessibility and market integration of hitherto isolated fragile areas into mainstream economy have also led to over-exploitation of CPRs (Jodha 1985a, b). As mentioned earlier, programmes like social forestry have changed the composition of products and poor people's access to CPRs (Chambers et al. 1989b). The argument here is not against the development of rural areas but against the underlying designs which failed to integrate CPRs into the development processes.

ROLE OF DONOR AGENCIES

The decline and degradation of CPRs may be attributed to the donor agencies only indirectly, in terms of their support to state policies that adversely affected CPRs, ranging from intensive agriculture programme and rural credit schemes to integrated rural development and watershed projects. On the face of it, this may seem unavoidable since donors operate through national agencies and that the latter alone should be held responsible for neglect of CPRs. But considering that when the donors insisted substantial amendments were made in the structure and content of different rural development projects, the donors are not entirely absolved of blame. The preconditions for several aid programmes, such as economic restructuring, environmental safety riders and, in some cases, even insistence on conditions relating to the choice of experts and source of material supplies for the projects, would bear this out.

There are not many circumstances which could prevent the incorporation of CPR concerns in donor-supported resource-centred programmes like drought proofing of drought-prone areas and watershed development project or social forestry projects in India. Agricultural R and D is another area in India's dry regions where donors have provided important support for technology development and have helped shape the direction of agricultural research. There too CPR perspective has largely been missing in the resource-centred research strategies.

The donors' disregard of CPR concerns may in fact be attributed to their operational style. First, since by circumstances the donors in the context of specific projects tend to operate on rather borrowed perspectives (i.e. perspectives of local agencies acceptable to the donors), they would not go beyond what their national counterparts would suggest. Secondly, even when the donors use their own perspectives, the latter too are usually uninformed about the institutional dynamics which is a core issue in the projects such as those involving CPRs. Thirdly, even when the donors have a deep understanding of the issues or desire to incorporate such issues in the projects, their appraisal missions or feasibility study groups operate under severe time constraints to meaningfully incorporate them in the projects. Finally, the CPR issues (as will be indicated shortly while discussing the role of other factors) have several micro-level dimensions. These are bound to be ignored in generalized schemes of interventions. The interventions need to be fairly location- or area-specific and much smaller in scale. This, however, may not fit with the donors' operational style. Consequently, relevant and potentially effective initiatives (e.g. the ones operated by NGOs) may not prove attractive to the donors due to 'economies of scale argument' given by them for not handling small-scale initiatives.

NON-UNIFORM IMPACT OF PUBLIC INTERVENTION

The foregoing discussion provides a generalized picture. The actual impact of public policies and programmes, however, would tend to vary, depending upon the local circumstances, including the local communities' resilience to changes forced from outside. Even within the same district, where the public policies influencing CPR were not different, inter-village differences in the decline of CPR area may be observed. Table 5.7 illustrates this by presenting inter-village differences in terms of (a) CPRs area as proportion of total village area during 1950–2 and 1982–4 and (b) percentage decline in CPR area during this period. Even within the same village the villagers may have varied approaches to different types and units of CPRs, indicating the differentiated village-level impact of uniform public policies. These differences could be part of the adaptation to the general collapse of traditional CPR-management systems. They could also be a product of village-specific circumstances, affected differently by uniform public policies. Hence the need to look into other factors influencing CPRs, with or without support of public policies. We might group these

Table 5.7. Range of Inter-village Differences in the Extent and Decline of CPR Area (percentage)

State (with no. of villages)	Range of CPR Area as Proportion of Total Village Area in		Range of Decline in the Area of CPRs
	1950–2	1982–4	1950–2 to 1982–4
Andhra Pradesh (10)	9–30	5–20	25–56
Gujarat (15)	7–31	2–23	21–69
Karnataka (12)	6–36	4–30	16–50
Madhya Pradesh (14)	29–69	19–47	14–51
Maharashtra (13)	8–43	6–34	14–52
Rajasthan (11)	20–49	8–26	17–71
Tamil Nadu (7)	7–39	5–23	21–65

Source: Adapted from Jodha 1986a, based on field work and village records.

factors as (a) demographic, (b) ecological, and (c) market related. For want of sufficient data on other aspects, we shall examine their impact only in terms of decline of CPR area, though mostly, there is not much divergence between the trends relating to decline in area, productivity and management of CPRs.

THE ROLE OF OTHER FACTORS: DEMOGRAPHY, ECOLOGY AND MARKET

The village-specific features considered for these three categories of factors influencing CPR area are as follows. Demographic factors include size and density of population; number of households; and growth of population during 1951–81. Some qualitative attributes of the population (e.g. occupational shifts, degree of factionalism and socio-economic differentiation, etc.) are also considered for a more detailed analysis. Ecological factors include area of village (the size of the village being often negatively associated with the harshness and marginality of agro-climatic environment); extent of sub-marginal lands (e.g. low fertility, gravelly, sandy, woody lands with undulating topography and some incidence of salinity, waterlogging, etc.), which are usually kept as CPRs; predominance of extensive pattern of land use reflected by the importance of livestock farming, etc. Market-related factors include distance from the market centre; proportion of cash crops in total cropped area; and extent of communication facilities conducive to market orientation of agriculture. An analysis of village-level data (through both bivariate tabulation and regression method) was attempted to see the association between the above variables and the extent and decline of CPR area. The broad picture revealed by the analysis (see Chart 5.3) is as follows.

(a) In smaller and isolated villages, where traditional social sanctions are still respected, the decline of CPR area is less. Transaction costs of enforcing social discipline regarding CPRs are lower in such cases. It is easier to organize 'user groups' and group action for CPRs in such a situation, as shown by the experience of different NGOs.

(b) In villages with relatively greater distance from market centres, where market forces are less effective in eroding traditional values *vis-à-vis* CPRs, the protection of CPR area is better.

(c) In smaller and isolated villages (often located in biophysically less favourable environments) ecological compulsions to retain and protect CPRs are stronger.

(d) In villages with smaller initial extent of CPRs, where communities have fuller knowledge and an active concern about their common resources, decline of CPRs is less. Informal social guarding of CPRs is easier in such areas.

(e) A further analysis of association between qualitative features of village population and the status of CPRs was attempted. Accordingly, the groups of villages with highest and lowest values of specific demographic characteristics (e.g. factionalism, occupational shifts, etc.) were compared with respect to decline in their CPR area. The details are presented in Table 5.8. A similar analysis was also done by comparing the groups of villages with highest and lowest decline of CPRs *vis-à-vis* the values of demographic characteristics mentioned above. This also confirmed the results indicated by Table 5.8, which are summarized below.

The decline in CPR areas is lower in the villages with the following characteristics:

(a) lower extent of occupational changes (i.e. shift from handicrafts, caste services, etc. to cultivation), implying lesser increase in the demand for conversion of CPR lands into private crop lands;

(b) lower degree of commercialization, implying lesser erosion of social sanctions and informal arrangements protecting CPRs;

(c) lower extent of factionalism in the village, implying greater degree of social cohesion, conducive to protection of CPRs;

(d) lower socio-economic differentiation ensuring equity of access

Chart 5.4. General Inferences from an Analysis of Village-level Data on Status and Changes in CPR Area and Associated Factors

Attributes of CPR Area Villages	Relative Position of CPRs			
	Extent of CPR Area (1950–52)		Decline During 1950–2 to 1982–4	
	Higher	Lower	Higher	Lower
Demographic Factors:				
Higher population		✓		
Greater no. of households		✓	✓	
Higher population increase overtime			✓	
Ecological Factors:				
Larger area of village	✓		✓	
Larger extent of sub-marginal lands	✓		✓	
Larger initial area of CPRs			✓	
Greater importance of livestock	✓			✓
Market-related Factors:				
Greater distance from market centre	✓			✓

Source: Based on comparison of groups of villages in each state having highest and lowest extent of CPRs (1950–2) and those having highest and lowest decline of CPRs (1950–2 to 1982–4). For details, see Jodha 1990a.

and benefits from CPRs, equal stake in the maintenance of CPRs and lesser extent of CPR grabbing;

(c) lesser dependence on state patronage for resource transfers to village, implying lesser opportunity for interference in village affairs from above, including for privatization of CPRs as part of populist programmes.

The above discussion of CPR changes and role of village-level factors, in a way depicts the whole process of rural transformation viewed from CPR perspective. Through the details under Chart 5.4 and Table 5.8 and their discussion, we have barely summarized the central role of the institutional dynamics affecting CPRs. The operation of this dynamics is further reflected through various actions and events (some of them to be summarized under Chart 5.5 later) that indicate people's relation to their natural resources and to each other. An understanding of these relationships can further explain the inter-village or inter-CPR-unit differences in the status and management of CPRs. Public intervention directed at group initiatives and

Table 5.8. Decline of CPR Area in the Villages Differentiated by
Qualitative Changes in their Population (1950–2 to 1982–4)

Changes in Demographic Characteristics	Villages Covered (no.)		Decline in CPRs Area (%)	
	Higher	Lower	Higher	Lower
Occupational Change: Proportion of household who newly shifted to agriculture	27[a]	21	37	12
Degree of Commercialization	31[b]	28	44	18
Factionalism, Group Dynamics	30[c]	26	28	14
Socio-economic differentiation	28[d]	32	42	15
Dependence on State Patronage	25[e]	27	60	16

[a]Households who shifted away from traditional caste occupations and became cultivators. Their proportion to total households in the village ranged from 15 to 20 per cent and 2 to 5 per cent respectively in the villages with 'higher' and 'lower' occupational shifts.

[b]Accessibility to market and related facilities are used as proxy for commercialization. Better accessibility is broadly defined to include the situation of villages having market centre within two kilometre of distance, availability of more than five shops in the village, regular operation of town-based trader or his agent in the village, year round bus service, etc. On the basis of the presence or absence of these attributes villages are grouped as those having 'higher' and 'lower' degree of commercialization.

[c]'Higher' factionalism means presence of two or more factions in a village with vast differences in their strength and political patronage from above. Villages with 'lower' factionalism lacked these features. They had factions of equal strength to be able to control each other.

[d]Differentiation reflected by values of Gini-coefficient of owned landholdings, which ranged from 0.63 to 0.75 and 0.34 to 0.40 in the villages with 'higher' and 'lower' socio-economic differentiation respectively.

[e]Villages which had officially sponsored land (*patta*) distribution functions more than twice during last ten years, had cent per cent dependence on state grant for CPR improvement, are put 'Higher' group. The villages with 'lower' patronage did not have these attributes.

Source: Adapted from Jodha 1988b.

management of natural resources at the micro-level can greatly benefit from the lessons offered by such understanding.

ADAPTATIONS TO THE CHANGING CPRs

Rural communities as the key actor in the field, operating under the influence of public interventions or pressures generated internally (e.g. through population growth), have contributed to the decline of CPRs at the village level. In the process they have also evolved their own strategies for coping with the changing CPR situation. The primary focus of such strategies is on maximization of private gains from the worsening status of CPRs. This, however, does not exclude small initiatives directed at the protection and rehabilitation of CPRs in some cases.

The extent and type of private gains extracted from rapidly declining CPRs being closely related to the capacities and needs of individual families, the adaptation strategies are shaped accordingly. The responses of the rural rich may differ noticeably from that of the poor towards the changing situation of CPRs, although some responses may be common. For instance, both attempt to grab CPR lands, with varying success. Chart 5.5 summarizes some relevant aspects of response strategies adopted by different groups in the study villages.

Changing CPRs and the Rural Rich

As observed during the field work and corroborated by other evidence, the dominant responses of the rural rich (e.g. large farmers) to the changing situation of CPRs include the following:

(a) Withdrawal from use of CPR products, as their opportunity cost of labour for collecting/using CPR products exceeds the value of CPR products.

(b) Increased reliance on alternative options. Alternatives include own supplies of biomass, substitution of renewable CPR products by non-renewables and/or external products (e.g. stone fencing for thorn fencing, or rubber tyres for wooden tyres for bullock-carts, iron tools for local-made wooden ones).

(c) Private squeeze on CPRs as assets, as reflected through a tendency to grab CPR lands, preventing others from using their private land during off-season (i.e. seasonal CPRs), and enriching own soil by mining silt and top soil from CPR lands to private fields.

(d) Indifference to management of CPRs despite their influence and ability to use legal-cum-administrative superstructure and public funds (grants/subsidies) available for rehabilitation of CPRs.

Perpetuation of the above responses would mean further decline in area and production from CPRs, and ultimately the complete irrelevance of community resources for the rural rich.

Chart 5.5. People's Adaptations to Changing Situation of CPRs[a]

Rural Rich	Rural Poor	Rural Community (General)
Withdrawal from CPRs as user of products Opportunity cost of labour being higher than CPR product value *Increased reliance on alternative options* Own bio-mass supplies; (stall feeding, etc.) Non-renewable/external resources (e.g. replacing stone fencing for thorn fencing, wooden tyres for local, wooden ones *Private squeez on CPRs as assets* Grabbing CPR land Preventing others using seasonal CPRs (private crop lands during off-season) *Approach to CPR management* Indifference to decline of CPRs As rural influential party to non-functioning legal and administrative superstructure for community resources	*Use of CPRs as an important source of sustenance* Complementarity of CPR–PPR based activities *Acceptance of inferior options* Opportunity cost of labour lower than value of products of degraded CPRs *Measure reflecting desperation* Premature harvesting of CPR products Removal of roots/base of products Over-crowding and over exploitation of CPRs Use of hiterto unusable inferior products	*Acceptance of CPRs as open access resources* Over-exploitation without users' obligations, regulations *Selective approach to specific CPR units* Despite general neglect of CPRs concern for some units *Focus on other users of CPRs* Item in seeking government subsidy/relief, in running factional quarrels, in populist programme, etc. *Part of non-operating legal and administrative measure* Changes in livestock composition (replacing cattle with sheep/goat, etc.) Agro-forestry initiative (revival of indigenous agro-forestry, etc.)

Source: Based on field work (1982–5). For elaboration and evidence, see Jodha 1989a; Iyenger 1988; Brara 1987; Arnold and Stewart 1991; Chambers et al. 1989b.

CHANGING CPRs AND THE RURAL POOR

Depending on their capacity, poor households also attempt some of the measures adopted by the rural rich. Their specific responses, however, include the following:

(a) Utilize CPRs as an important source of sustenance and attempt maximization of complementarity between CPRs and PPRs (Tables 5.3, 5.4).
(b) Ready acceptance of increasingly inferior options offered by CPRs. This is both because of non-availability of alternative options and the falling rate of opportunity cost of labour with their increasing poverty in most cases.
(c) Resort to measures manifesting a high degree of desperation. Examples are: increased frequency and unseasonal lopping of trees and premature harvest (collection) of CPR products, reducing seed formation and regeneration possibilities; removal of plant/bush roots (the very basis of CPR products); use of hitherto discarded (inferior) products with negative side effects on the health of the users; and overcrowding and over-exploitation of CPRs.

The consequence of these trends will be further degradation of CPRs and overall environmental resources at the village level, and rapid decline of whatever cushion the rural poor have through CPRs.

CHANGING CPRs AND THE RURAL COMMUNITY

A number of response measures against the changing CPR situation adopted by different groups in the villages are reflected clearly at the total community level. They include the following:

(a) General acceptance of CPRs as open access resources, following the abolition or disintegration of traditional usage regulations. This is reflected by complete absence of users' obligations and consequent over-exploitation of CPRs on the one hand and failure to question the non-functional legal and administrative measures relating to CPRs.
(b) Focus on alternative uses or rather misuses of CPRs, as reflected by treating CPRs as an issue in factional disputes, projecting CPRs as an item to secure government subsidies/grants for village panchayats, using CPR lands for distribution of political patronage, etc.

(c) Structural changes, to be elaborated shortly, as reflected through change in the composition of livestock (Table 5.9), revival of indigenous agro-forestry, and search for alternative options, etc.

(d) General neglect of CPRs with selective management approach to specific CPR types and units (as revealed by Table 5.10), to be elaborated shortly.

The last two categories of community responses to the changing CPR situation are elaborated further.

CHANGES IN THE COMPOSITION OF LIVESTOCK

Since CPRs are a mainstay of the community's livestock, their decline has forced significant adjustments in livestock management. Reduced grazing space and depletion of forage potential have brought about both a reduction in the size of animal holdings and changes in their composition. Indications to this effect, given by detailed studies based on data at two points of time in the villages of Rajasthan (Jodha 1985b) are further confirmed by similar enquiries largely based on oral histories and current position of sample households in other areas. It is seen from Table 5.9 that the number of bullocks among the sample households declined by 19 to 42 per cent in different areas between 1950–2 and 1982–4.

Table 5.9. Structural Changes in Livestock Population in Response to Decline of CPRs in Selected Villages

State (with no. of villages and sample households)	Percentage Change in the Number of Animals between 1950–2 and 1982–4[a]			
	Bullock	Cow	Buffalo	Sheep & Goat
Andhra Pradesh (2, 38)	- 21	- 18	+ 4	+ 23
Gujarat (4, 68)	- 30	- 26	+ 20	+ 19
Karnataka (2, 40)	- 22	- 19	+ 9	+ 22
Madhya Pradesh (4, 80)	- 16	- 18	+ 12	+19
Maharashtra (4, 82)	- 31	- 19	+ 12	+ 32
Rajasthan (6, 115)	- 42 (63)[b]	- 35	+ 14	+ 38
Tamil Nadu (2, 30)	- 19	- 14	+ 9	+ 21

[a]Based on current status and oral history of animal holdings (in 1950–2) of sample households recorded during the field work during 1982–5. Data for the past adjusted for division of families, etc.

[b]Figures in parentheses indicates change in the number of camels in Rajasthan villages.

Source: Adapted from Jodha 1990a.

Table 5.10. Factors Inducing the Adoption of Management Measures for
        CPR units in Selected Villages

Factors Underlying the Adoption of Measures	Percentage Distribution of Management Events According to the Underlying Factors			
	Area Protection (474)	Usage Regulation (423)	Development (532)	Total (1429)
A. *CPR-unit-related factors*				
1. High productivity, visible contribution to private farming	12	12	13	12
2. Location, size, proximity to village	12	9	8	10
3. Usability for seeking government grant	15	5	20	14
B. *User-related factors*				
1. Private stake/control of influential groups	5	12	10	8
2. Short-lived provocation against irregularities, encroachments, etc.	8	11	-	6
C. *By-product of other activities*				
1. Factional politics of village	24	25	8	19
2. Rituals/religions sanctions	5	9	3	4
3. Provisions under development/relief programmes	13	7	23	17
D. *Genuine/positive factors*				
1. Concern against degradation, irregularities; enlightened leadership, NGO activities	6	10	15	10

Notes: The figures in parentheses indicate total number of measures adopted. Percentage distribution of these measures according to the underlying factors is presented in the respective columns.

These factors can be grouped under alternative categories such as: (A. 1-3, B-3) Economic; (B. 1-2, C. 1-3) Political; (C-2) Religious; (D-1) Environmental concerns and participatory processes.

Source: Case histories of 176 CPR units. See Jodha 1990a for details. Also see Annexures 5.2a to 5.2c.

Maintaining unproductive animals without CPR support is difficult. The reduction in the number of bullocks is partly explained by increased mechanization in certain areas, and partly by the high overhead cost of maintaining bullocks. Feeding the bullocks for the whole year without CPR support becomes uneconomical when they are in agricultural use for only three or four months. The number of cows also declined for similar reasons. In dry areas, owing to droughts and frequent miscarriages, etc. the cows' unproductive period often exceeds their lactation period. Also, the high private cost implied in increased stall feeding favoured buffalo keeping as against cows. Favouring the buffalo is also fat-based milk pricing and the buffalo's higher milk yield.

An even more significant change in livestock is the substantial increase in the goat and sheep population. Not only could the small ruminants be sustained by degraded CPRs, they fit better in the changed migration patterns. Reduced CPRs favour the migration of small ruminants over that of cattle. Thus sheep and goat, often accused of destroying CPR vegetation, seem to have attained importance, following the degradation of CPRs. Broadly similar changes in the composition of livestock have been recorded by other micro-level studies (Iyengar 1988; Ahuja and Rathore 1988).

SELECTIVE APPROACH TO CPRs

While generally neglecting CPRs the villagers may have a selective approach to specific units of different CPRs. A closer understanding of this phenomenon can offer useful leads for future strategies for development and sustained use of CPRs. A detailed enquiry of over 175 units of different CPRs from different villages revealed the emerging patterns of CPR management. The term management is defined to cover people's (as against government's) interventions for (a) area protection, (b) usage regulation, and (c) development or upkeep of CPRs. Based on detailed case histories (covering a period of thirty to forty years) of 176 CPR units, an inventory of nearly 1450 events involving 'people's interventions' in CPR matters, as indicated above, was prepared. The distribution of such events (i.e. CPR management events), according to the factors inducing them, is presented in Table 5.10. More details are summarized under Annexures 5.2 to 5.4. The important inferences suggested by the table are as follows:

(a) The bulk of the CPR-unit-specific management events are a by-product of other developments, such as factional quarrels in the village or adherence to certain rituals and religious sanctions or specific conditions of government grants to the village. For instance if the area of any CPR helps villages to qualify for certain grant or relief, they try to keep its area intact even without developing it or regulating its use. From this perspective, management or the future of CPRs is tied to their usability for other purposes rather than their utility as community assets. While factionalism and rituals (unless used in some innovative ways) offer little usable leads for CPR strategies, widening of criteria for government grant based on CPRs to include their management and productivity, can be of great help.

(b) Higher productivity and yield of CPRs play an important role in inducing their better management. This becomes more important when these gains are shared more equally. The productivity–management linkage offers a useful clue for breaking the vicious circle of 'degradation–neglect–degradation' characterizing CPRs.

(c) CPR unit's location (e.g. in the context of a watershed), size, and proximity to village, as well as rituals and religious sanctions affecting specific CPRs, also play a positive role in the management of CPRs. Since most of these factors cannot be easily manipulated, in general they may not provide any operational basis for forward-looking practical policies for CPRs. Yet they suggest a need for discriminating and location-specific measures to develop and manage CPRs with greater involvement of the local people.

(d) A genuine concern against degradation and misuse of CPRs is an important factor inducing people's action for CPRs. This accounts for the small proportion of management events in the study villages, but this offers a potentially viable option for rehabilitation and better management of CPRs. With the involvement of NGOs and enlight-ened villagers this can be further strengthened.

People's discriminating approach to different CPRs/CPR units further confirms the role of local-level, institutional factors in the management and upkeep of community resources. However, they also offer useful leads for macro-level policies which can facilitate at the village level better management and efficient use of not only CPRs, but other interventions involving participatory processes. Their important lessons for CPR strategies may include: (a) making CPR management an explicit part of the criteria for resource transfers to villages; (b) focus on productivity-promoting measures to break the nexus between low productivity and neglect/degradation of CPRs;

(c) differentiating the application of generalized CPR policies according to specific characteristics of CPR units at the village level (e.g. village forest plots with differences in size, physical accessibility, level of productivity, etc. to get different treatments); (d) active policies and programmes encouraging NGOs and others to evolve village-specific measures for CPRs and protect their initiatives from legalities of generalized state policies.

PROSPECTS: WHITHER CPRs?

The evidence and discussion presented above do not suggest bright prospects for CPRs in India's dry regions. The major factors constraining the present and future of CPRs, as revealed by the above discussion, are recapitulated in Chart 5.6. The chart also summarizes the factors which justify the rehabilitation of CPRs and possible steps to do so.

CONSTRAINING FRAMEWORK

According to Chart 5.6, the institutional framework directly related to CPRs is highly constraining. Undeclared regressive state policies, encouraging privatization and neglect of CPRs are the primary factors causing rapid decline of CPRs. Both physical and legal-cum-administrative interventions dealing with CPRs are insensitive to CPR perspective. The rural peoples response to the changing CPR situation is dominated by a tendency to grab CPR area and over-exploit their production potential. Finally, there is neither a users' lobby nor a noise-making media to plead for CPRs.

RATIONALIZATION OF CPR DECLINE

Besides these factors, a number of circumstances associated with the current position of CPRs are often interpreted to rationalize their decline. The key arguments come in four guises, as follows:

(a) *Efficiency Argument:* Privatization of CPRs is suggested as a possible solution for their physical rehabilitation and sustained productivity. This is a part of the often repeated recommendation for handling the so-called tragedy of commons. Recent evidence from different parts of the world, however, suggests otherwise (Runge 1981; Bromley and Cernea 1989; Repetto and Holmes 1984; Ostrom 1988; Feeny et al. 1990). There is insufficient evidence for fully analysing this issue in the context of India's dry regions.

Table 5.11. Distribution of Privatized CPR Plots by Their Comparative
Production Performance[a]

State (no. of villages and ex-CPR plots)	Proportion of Privatized CPR Area Transferred to Annual Cropping[b] (%)	% Distribution of (i) ex-CPR Plots According to Their Yields as Proportion of Yield of (ii) Plots Traditionally Cropped [Yield of (ii)=100] %[c]			
		10–25	26–50	51–75	76+
Andhra Pradesh (1,65)	96	15	43	25	17
Gujarat (2,90)	82	11	50	20	19
Maharashtra (2,85)	93	17	61	7	15
Madhya Pradesh (2,98)	78	12	39	19	30

[a]Based on plot-wise details collected under ICRISAT's village level studies; average of two cropping years (1983 and 1984).
[b]This information relates to total CPR land privatized in the village rather than selected ex-CPR plots for which yield was recorded.
[c]Comparison is based on observations generated by the following procedure. For each (i) ex-CPR plot another, (ii) plot traditionally cropped belonging to the same farmer and put under the same crop was picked up. Grain yield of (i) as proportion of yield of (ii), constituted one observation.
Source: Adapted from Jodha 1990a.

Nevertheless, the prevailing situation could be described as follows. CPRs in dry regions largely consist of sub-marginal and fragile land areas. Efficiency of their use should be seen in the context of their use capabilities. Accordingly, their retention under natural vegetation (e.g. through CPR) is an effective approach for their efficient use.

The facts as presented in Table 5.11, however, indicate otherwise. In different villages 78 to 96 per cent of these sub-marginal lands were transferred to annual cropping following their privatization. Their crop productivity (as compared to the prime lands traditionally cropped) is very low. Nearly half of the plots of formerly CPR land had grain yields 50 per cent or less than in the prime land plots cultivated by the same sets of farmers. There are no data to reflect on the productivity of these plots prior to their privatization. Limited evidence based on detailed case studies of plots belonging to selected farmers in the study villages from Rajasthan and Gujarat is presented in Table 5.12. According to the table grain yield of former CPR plots fell far short of yield from traditionally cultivated plots. Fodder and fuel (biomass) production did, however, increase substantially

Table 5.12. Impact of Privatization on Biomass Productivity of CPR Plots: Production Performance

	CPR Plots	Former CPR Plots[a]		Traditionally Cropped Plots
		Kept under Natural	Put under Crops	
Plots (no.)	6 (8)[b]	4 (5)	12 (16)	13 (16)
Capital investment (Rs)[c]	Nil	300 (428)	1200 (1530)	300 (700)
Fodder/fuel collection (cart-load/ha.)[d]	2 (3)	8 (10)	-	-
Animal units grazed (no./ha. per day)[e]	46 (34)	7 (11)	-	-
Grain production (qtl.)[f]	-	-	168 (203)	425 (519)
Beneficiary households (no.)	112 (81)	8 (10)	6 (7)	6(7)

[a]Area of plots ranged between one and six hectares. Plots under cols 2 to 4 had similar soil conditions. Plots under cols 3 to 5 belonged to the same farmers.
[b]Based on case studies of selected plots in one village each in Rajasthan and Gujarat, figures for the latter being in parentheses. Average for two cropping years (1983 and 1984).
[c]Expenditure on permanent improvements, e.g. fencing, ridiging, trenching during the last five years.
[d]A cart-load of biomass weighing approximately 5 quintals.
[e]During four months of rainy season.
[f]Pearl millet (bajra) yield.
Source: Adapted from Jodha 1990a.

Chart 5.6. Factors Influencing Prospects of CPRs

Constraining Framework	Imperatives Supporting Rehabilitation	Future of CPRs: Possible Options and Dilemmas
Undeclared, regressive state policy towards CPRs (privatization, lack of management).	Ecological, environmental and long-term sustainability concerns (i.e. required resource use systems in regions with sub-marginal lands and high climate variability).	Positive policies restricting further reduction in CPR area (Obstacle: new social culture—collective indifference and land grabbing).
People's response: land grabing, over-exploitation, and indifference to CPRs.	Complementarity of CPR–PPR-based farming systems (i.e. due to non-covariability of input needs and product flows and narrow and unstable base of private crop farming).	High investment needs for high productivity (Obstacles: long gestation period, invisibility of gains by narrow cost-benefit norms).
Missing CPR-perspective of development interventions (fiscal, technological and institutional measures for CPRs).	Sustenance of rural poor (through product supply, employment and income generation, etc.).	Rehabilitation and sustaining of CPRs as high-productivity community assets; (Technology with focus on diversification and user perspective; management by user groups based on collective stake and equal share in gains).
Negative side effects of development/ transformation processes (commercialization, etc.).	Opportunity for evolving participatory development approaches.	
CPRs made open access resources, conducive to tragedy of commons.		

(by three to four times) in the former CPR plots when retained under natural vegetation.

Table 5.12, however, reveals two additional factors associated with the increased biomass productivity of former CPR plots. First, the improved production performance is a result of capital investment, protection and restricted use (indicated by animal units grazed). If the same magnitude of investment and low-usage intensity are applied to CPRs, their performance too can be upgraded significantly, as different studies (Shankaranarayan and Kalla 1985; Oza 1989) indicate. Secondly, quite obviously, privatization has restricted the use and access facilities of former CPRs to a very small number of animals and households. This raises the basic equity issue, where higher production prospects for a few result in loss of (low-productivity) options for many. Moreover, given the limited availability of CPR lands, it will be impossible to help even a fraction of the rural population by distributing CPR lands to them. Even if all CPR lands were to be distributed, in 87 per cent of the study villages, not more than 12 to 20 per cent of the poor households would get more than one hectare per family.

(b) *'Poverty' Argument:* It is argued that since the decline of CPRs is only a manifestation of the pauperization process in the country, the solution to the CPRs problem lies in eradicating rural poverty and land hunger. In answer, without belittling the significance of linkages between poverty and CPR depletion, it needs to be pointed out that the above argument implies perpetuating the vicious circle implied by the above linkages. The solution to poverty does not lie in further impoverishment of CPRs; rather, it would partly lie in their rehabilitation and regulated use. The noted difference between employment and income from CPR-based activities in villages with better managed CPRs and the others would support this.

(c) *'Inevitability' Argument:* Proponents of this view argue that, as indicated by Chart 5.6 and confirmed by the experience of developed countries as well as agriculturally developed pockets within India's dry regions (Iyengar 1988), decline of CPRs is a part of the development process. Transformation of rural areas contributes to the erosion of CPRs. In answer, it needs to be pointed out that unless the development process also reduces people's direct dependence on the CPRs, the decline of CPRs would mean depriving the people of various services and products offered by CPRs. Viewed this way, the utility and relevance of CPRs is undiminished in Indias dry regions at their present stage of development.

Furthermore, instead of their deliberate marginalization, if CPRs are incorporated as an integral part of development, the consequent enhanced contribution of CRPs to rural transformation would further question the 'inevitability argument'.

(d) *'Scarcity' Argument:* Related to the 'inevitability argument' is the reasoning based on land scarcity. In the context of rising pressure on land, CPRs are considered as the only option to satisfy mounting land hunger through legal or extra-legal means to privatize them. In answer, it needs to be pointed out that population-induced land scarcity does not seem to be a sufficient condition for decline of CPR area. At least the evidence from the present study shows groups of villages with high population increase and yet limited decline in CPR area and vice versa.

Furthermore, the ultimate concern of public policies should be the elimination of human hunger and scarcity of agricultural products rather than mere satisfaction of land hunger. The fact that poverty and scarcity persist for the families even after acquiring CPR land as private property further reduces the validity of CPR privatization as an answer to the scarcity problem. On the contrary, specific contributions of the CPRs in terms of sustainable supplies of biomass and stability of farming systems may be permanently lost and thus accentuate the problem of hunger, once CPRs are privatized and invariably converted into low-productivity crop lands. The environmental impact of the disappearance of CPRs is another weighty consideration.

Most importantly, grabbing and privatization of CPRs as well as their over-extraction, as elaborated earlier, are in a way forms of forced adjustments to declining CPR situation. They cannot be a substitute for 'anticipatory measures' which, rather than marginalizing and in the process eliminating CPRs, can develop and harness CPRs as productive social assets. Thus the 'scarcity argument' seems to derive its validity from 'default' on the part of the policy-makers to have a positive approach to CPRs.

POSITIVE CONSIDERATIONS

Arguments condoning the decline of CPRs, thus have insufficient validity. There are on the other hand, positive considerations supporting the case for CPRs. These relate to the functions and services of CPRs as indicated by Tables 5.1–5.4. Accordingly, in the approach to CPRs the central concerns ought to be (a) ecological and environmental imperatives of the natural resource base of the dry regions, (b) CPR–PPR complementarities, and (c) sustenance needs of the rural poor. There is yet another more fundamental issue which relates to (d) the rising concern for incorporating grassroot democracy and traditional wisdom into the conventional development culture. CPRs offer a potential area for evolving and implementing options and approaches in this regard.

(a) *Ecological and Environmental Imperatives:* Both heterogeneity of land resources and highly variable climate conditions, call for diversified resource use and keeping of sub-marginal/fragile lands under low-intensity uses (e.g. natural vegetation as against annual cropping). Provision of CPRs helps satisfy these requirements. The same goal can be achieved if privatized CPRs are retained under natural vegetation. As shown in Table 5.11, however, where 78 to 96 of sub-marginal lands are shifted to crops following their privatization, PPRs performing the CPRs' ecological function does not seem likely.

Furthermore, the stability and productivity of environmental resources in the dry-land context are greatly influenced by the way fragile resources (e.g. CPRs) are managed—and protected (Jodha 1991a). Loss of CPRs would mean the loss of an important means of handling the environmental problem in dry areas.

(b) *CPR-PPR Complementarity:* Related to the ecological issue, CPRs (due to a different production cycle of their natural vegetation) have input needs and output flows which are qualitatively and temporally different from those of PPR-based crop farming. This forms the basis of complementarities of production systems based on CPRs and PPRs (see Table 5.2). To the extent natural vegetation on CPRs facilitates the above complementarities, keeping part of private land under natural vegetation can perform this function. But it is not the natural vegetation alone, but accessibility to it, which is responsible for CPRs serving as a cushion when PPRs fail to meet the needs (see Chart 5.5). In such a situation, there are no ready alternatives to CPRs, for maintaining and strengthening the PPR–CPR type complementarities and ensuring associated benefits, especially in the high-risk environment of dry tropical areas.

(c) *Sustenance of the Rural Poor:* The more pressing requirement is the contribution of CPRs towards the sustenance of the rural poor, who lack alternative options (see Table 5.1). Notwithstanding a number of measures initiated to help them, there are not many programmes which can match the CPRs in helping them (Jodha 1986a). Enhanced productivity and their regulated use can go a long way in raising the contribution of CPRs to the quality of life of the rural poor. The cost of abolishing CPRs, in terms of forgone opportunities for collective gains to the poor, would be too high to be compensated by other means.

(d) *Changing Development Paradigm:* The present essay, though focused on CPRs, has raised several basic issues relating to the 'conventional development culture', that characterizes public intervention. A few of its relevant features include: greater emphasis on information as against understanding, central place to technology at the cost of institutional factors, state or formal agencies as key actors reducing user groups and people into marginal entities, etc. There is an increasing concern against the side-effects of these tendencies. The need for the incorporation of local concerns, gender issues, participatory development, and sensitivity to people's perceptions and traditional wisdom are increasingly voiced. CPRs offer an ideal field to test these concerns and evolve options for wider use in the development programmes involving people's participation, local management of local resources, etc. Some of these issues are discussed below.

PROMOTING CPR CAUSE: ADVOCACY AND ACTION

In this concluding section we largely summarize the major highlights of the present discussion with focus on their policy and programme implications. We also allude to the major actors in the game, including donor groups, who can make a significant contribution towards the rehabilitation and sustained use of CPRs for the betterment of the rural communities.

The Basis for Hope

The case for CPR development is sustained by the varied contributions of these resources rather than by any sentiments for collective resources. Without being romantic about traditional resource management systems and CPRs as a manifestation of the same, some objective realities may be stated, which can make a strong enough case for the promotion of CPRs as a part of rural development strategies, particularly in India's dry tropical regions.

First, the visible and invisible gains from CPRs, as reported in this essay and other studies, far exceed the conceivable disadvantages associated with CPRs. The so-called tragedy of commons becomes operative when CPRs are converted into open-access resource through default on the part of public policy (Bromley and Cernea 1989; Feeny et al. 1990).

Secondly, owing to the CPRs' contributions towards the stability and sustainability of farming systems, sustenance of the rural poor, management of local resource by local people, environmental stability in the village context, etc., the CPR-centred policies and programmes would have a strong convergence with the policy focus of several development strategies currently promoted by developing countries and donors alike. The examples may include a whole range of anti-poverty programmes in rural areas, measures directed at ecological and environmental stability, strategies for encouraging participatory development and sustainable resource use, etc.

Thirdly, most of the factors and processes contributing to the decline of CPRs can be controlled through appropriate changes in public policies and other circumstances affecting CPRs. This forms the basis of advocacy to protect and develop CPRs for their sustainable use.

Finally, the world community (including the developing countries) has accumulated sizeable evidence on successful initiatives on the management of common resources through community involvement for equitable gains. This offers hope for redesigning management systems for CPRs.

The preceding two points form the immediate context for the following discussion on the rehabilitation and development of CPRs.

KEY AREAS OF ATTENTION

The key areas where awareness, advocacy and action initiatives could be focused for CPR development are: public policies, technology and investment issues; and management of CPRs through effective involvement of CPR users. They are elaborated below.

(a) *CPRs and Public Policy:* As stated earlier, a primary reason for the decline of CPRs in India's dry regions is the indifference of public policy towards these resources. To alter the situation the policy environment needs to change. This could happen in three directions, as indicated below.

(i) Positive CPR Policy: This implies the need for active policy and programme to restrict further curtailment of CPR areas; regulation of use intensity of these resources through appropriate usage policy; provision of punishment for violations; empowering people (e.g. user groups) to manage resources, etc.

(ii) Side-effects of Development Intervention: Various welfare and development interventions are undertaken without evaluating their potential impact on CPRs. The programmes ranging from land reform to subsidies for agricultural mechanization come in this category, which adversely affect CPRs. The provision of some 'rider' in terms of the CPR sensitivity of a project (like environmental sensitivity) can go a long way in safeguarding CPRs and their productivity.

(iii) General Development Policies: The current policies and programmes relating to integrated rural development or environmental protection or poverty eradication, etc., contain a number of elements which could be more effectively implemented with focus on CPRs as a project component. This could be done on the basis of a proper understanding of the potential contributions of CPRs.

While suggesting these policy approaches, however, one should be aware of the circumstances which may obstruct the initiation and implementation of such policies. In the Indian context the policy-makers' high propensity for 'populist programmes' may prove a key obstacle, as the distribution of CPR land to the people has always been used as a means of pleasing the people. Similarly, the minimization of the side-effects of development interventions on CPRs, and integration of CPRs as focus areas in other development projects may be obstructed by the general ignorance and indifference of programme planners. The persistent lack of CPR perspective on the part of functionaries dealing with community resources, and emerging 'social culture', that has generated collective indifference towards CPRs and strengthened individual tendencies to grab or over-exploit them, are two other key hurdles in promoting the CPR cause. One possibility to guard against these obstacles is the development of a strong CPR lobby through the media and NGOs. This could be supplemented with policies directed to the promotion of CPR user groups. More on this later.

(b) *Investment Needs:* For sustained and effective contribution of CPRs, increase in their productivity is essential. This requires rapid regeneration (through protection and regulated use) and provision of substantial investment into CPRs. To convert CPRs from natural resources for extraction only into managed productive assets, planned investment is unavoidable. The key obstacles to higher resource investment in CPRs may include: (i) absence of fiscal tradition to patronize such community resources; (ii) long gestation period and a complex of transaction costs associated with resource allocation to CPRs; and (iii) invisibility of gains. The solution to these problems may lie in a deliberate decision on resource transfer to CPRs and widening of the narrow focus of investment yardsticks. Furthermore, increased pressure of users following the improvement in CPR is another possibility which, unless checked, can restart the process of resource degradation. The solution lies in the organization of effective user groups. In the investment strategies for the rehabilitation of CPRs, donor agencies can play an important role. However, to do this effectively, they may have to incorporate a CPR perspective in their approach to natural resource development.

(c) *Technology Focus:* It hardly needs reiteration that the present degradation of CPRs is partly due to the operation of the vicious circle involving degradation–neglect–further degradation. As discussed earlier, people can be induced to change their approach to CPRs if the CPRs are made more productive. To break the above vicious circle new technologies which can enhance regeneration, increase the flow of biomass, etc., are an important requirement. The rehabilitation of CPRs as productive social assets needs a new technological focus in terms of species, inputs and technical methods of resource management, etc. Besides productivity, the diversity and usability of products need emphasis. The key obstacles to this possibility are: (i) persistent gaps between the perspectives of technologists and resource users; (ii) inability to screen available resource-centred technologies for their institutional acceptability; and (iii) frequent high priorities to commercial considerations while designing technologies for community lands (as in the case of social forestry programmes). The remedial measures under such circumstances should start with the sensitization of scientists and R and D planners to the CPR perspective. Some work, already initiated under the integrated watershed development projects, has helped scientists to reorient their technologies to suit CPRs (World Bank 1989; Arnold and Stewart 1991).

(d) *Management and Regulation:* In a way, the rehabilitation of CPRs is less of an investment-cum-technological problem and more of a resource management problem. The impact of investment and technology may prove short-lived unless the management and usage aspects of CPRs are effectively handled. In most areas, even natural regeneration itself can make CPRs more productive, provided it is permitted through controlled and regulated use of resources. However, this cannot happen unless CPRs are reconverted from open access resources to common property resources. In operational terms this would mean the re-establishment and enforcement of usage regulations and user-obligations towards CPRs (Jodha 1985a, b).

At aggregate policy level this could be facilitated by provisions which not only give legal sanction to CPR usage practices but empower local communities to implement such provisions. Some of these provisions in terms of mandate to village panchayats do exist, but as mentioned earlier, panchayats have failed to implement them. One reason for this failure is the legal and formal status of panchayats which makes them a miniature replica of state authority rather than a representative body of CPR users. Consequently, village panchayats have failed to replace the traditional management system of CPRs, and common property resources became open access resources. A redeeming feature of the current situation, however (as revealed by the inter-village and inter-CPR unit differences in people's approach to CPRs) is that there are still certain elements which could be integrated into workable strategy for CPR management. The focal point of such strategies could be the organization of CPR-user groups.

(e) *CPR-User Groups:* The idea of CPR-user group in a way recommends itself. First, as stated earlier this fits well into the increasingly emphasized grassroot-level democratization of resource management systems and participatory development processes. Secondly, this could be an important approach to reducing the cost of policing and subsidizing resources and facilitating local resource mobilization. Thirdly, it has some equity-oriented elements. These positive factors could, however, be easily counter-balanced by just one factor, i.e. the difficulty in creating user groups. Left to the legal and administrative capabilities of the state, many more superstructures (like village panchayats) can be easily created, but they will be of limited use. The creation of genuine user groups calls for close understanding of various social and cultural features of village communities and their response strategies to new forces of change. The size of the group, its operational integrity, approach to internal equity, etc. are issues which cannot be imposed through a generalized scheme of promoting CPR-user groups. The groups have to develop in keeping with the local socio-economic and CPR-related circumstances. Without imposing specific models the state policies can facilitate this task by providing legal flexibility and logistic support for agencies like NGOs who, with their better feel of the field and close association with different groups of village communities, can help organize locally suited CPR-user groups.

There are no unique models to pattern such groupings in dry areas. The choice of the key characteristics of CPR-user groups can be based on some understanding of traditional forms of rural cooperation, a few insights revealed by the emerging patterns of CPR management (Table 5.10), and the experience of a number of successful initiatives tried for management of community resources in different parts of the country (Mishra and Sarin 1987; Chopra et al. 1990; Shah 1987; Agarwal and Narain 1990a; Poffenberger and McGean 1996). Experience from other developing countries (Bromley and Cernea 1989) can also help.

On the basis of the foregoing, we indicate some features of prospective CPR-user groups (for India's dry regions):

(i) The foremost attribute of a CPR-user group should be equity of access and benefits from CPR for all members.

(ii) CPR-user groups should have legal sanction, but they should be outside the control of formal institutions such as village panchayats, government's revenue department, etc.

(iii) Depending on the type of CPR and village-specific circumstances, membership of the group may comprise the entire village community or specific occupational groups.

(iv) Preconditions for group membership (besides being resident of the village and user of CPR) should include binding commitment to user obligations and usage regulations.

(v) To ensure stability of user groups flexibility in terms of exit and entry of members may be allowed with no right to break the group.

(vi) The provision of CPR-user groups can be viewed as an intermediate arrangement in between complete privatization and current communal usage system (implying open-access resource regime). The arrangements, however, should relate to access and usages, without any claim to resource itself.

(vii) Except for incorporation of the broad features like the above, CPR-user groups need not have a uniform pattern all over the dry regions or even throughout a single state. Depending on the type of CPR and village-specific circumstances the pattern may vary and evolve.

In the context of some dominant features of the current situation, these suggestions may sound utopian. The two relevant features which have emerged as the by-products of India's recent development history (and developing countries in general, and which may obstruct the growth of user groups are: (a) the ever-increasing tendency of the state to expropriate the initiatives and activities which belong to people; (b) increased internal differentiation of rural communities and its impact on the operation of village-level initiatives. Despite such potential obstructions, the success of recent initiatives in the management of community resources by user groups and NGOs do inspire some hope. Besides, the emerging awareness and grassroots-level pressure for local control of local resources and associating people in protecting their immediate environmental resources may also lend some strength to the case for CPR-user groups.

(f) *Role of Donor Agencies:* In the broad strategies to rehabilitate CPRs (including public policies, technological and investment support and management through user groups), the key actors are policy-makers, grassroot-level workers and the CPR users themselves. The policy-makers include both national decision-makers at different levels as well as the donor-agency functionaries, who may directly or indirectly influence the decisions of national policy-makers. In the context of CPRs, a brief reference to the role or donor agencies has already been made. The same could be summarized here.

The record shows that donor agencies have been as indifferent to CPRs in India as the national agencies. Despite evidence of the donor agencies' ability to influence the policies of national agencies in various sectors, the influence has not been felt in the case of CPRs.

Furthermore, in their substantial support to programmes involving natural resource development (e.g. community pastures, social forestry, etc.), the focus has been on techniques and funding rather than on resource users. These gaps are increasingly being realized by donor agencies themselves, as revealed by several evaluation exercises (Bromley and Cernea 1989; World Bank 1989; Arnold and Stewart 1991).

Besides the realization of these gaps, a few additional factors may facilitate the evolution of active CPR policies on the part of major donor agencies like the World Bank. They would include the emerging concerns such as: local management of local resources and participatory development; environment-friendly pattern of natural resource use involving people's participation at the grassroot level; cost-effective, self-sustaining approaches to poverty eradication, etc. The above concerns at donor agency level are complemented by similar concerns emphasized by the national agencies (though at times fluctuating with the degree of economic and political perturbations faced by them). In India, initiatives by agencies such as the National Wasteland Development Board (SPWD 1991), Planning Commission and several NGO groups (Shah 1987; Poffenberger and McGean 1996), strengthened by the evidence provided by field studies such as the one reported here, have already attempted to incorporate CPR perspectives into development interventions.

The more concrete areas where donor agencies could play an important role are (i) sensitizing national agency policies *vis-à-vis* CPRs (by adding a CPR-rider to relevant programmes); (ii) support to R and D and technology relevant to the CPR context; (iii) investment facilities focused on CPR development; (iv) support to institutional measures such as promotion of CPR-user groups and relevant NGO activities.

There are two important constraints to the donor agencies' positive support to CPRs. First, as in the case of national agencies, unless donor group representatives are sensitive to the CPR perspective, the case for CPRs will not get adequate and effective projection in their (feasibility, evaluation) mission reports. This is largely a matter of professional composition of missions and their terms of reference, which could be handled once higher-level policy issues are clear.

The second problem of donors' approach to CPRs relates to the possible incompatibility between the scale of large donors' initiatives and diversified and smaller scale of CPR activities. By the very nature of their operational style the large donors may find CPR cases too small and uneconomic to handle or process. One approach to such structural problem could be the provision of sufficient flexibility in large grants at the operational stage. Another approach may include some forms of subcontracting jobs to small agencies with direct links with small-scale activities. These approaches have already been tried by some donor agencies. They could be effectively extended to diversified requirements of CPRs.

Annexure 5.1. Location and Agro-climatic Features of the Areas Covered by the CPR Study[a]

State/District	Taluka/Tehsil	No. of Villages	Rainfall[b] mm	Soil Type[c]	% of CPRs to Total Village Area	Persons/Km² of Total Area in Taluka[d]	in the Village	Persons/10 ha of Area of CPRs in the Villages
Andhra Pradesh								
Anantpur	Anantpur	2	563	A-L,G	15	90	106	71
Mahbubnagar	Athmakur, Kalvakurthi, Mahbubnagar	5	721	A-D, A-L	9	124	162	186
Medak	Medak	3	834	A-D, V-D	11	177	158	145
Gujarat								
Banaskantha	Kankrej	5	655	S	10	210	201	205
Mehsana	Vadgam	5	633	S	11	340	332	301
Sabarkantha	Sidhpur, Vijapur Prantij	5	739	S,V-M	12	238	253	2085
Karnataka								
Bidar	Bhalki, Bidar, Humnabad	3	907	A-M, A-D	12	162	137	113
Dharwar	Kaighatgi	3	691	A-M,	10	134	156	154
Gulbarga	Gulbarga	3	702	V-M, A-D	9	106	129	146
Mysore	Gundulupet	3	680	A-M	18	106	103	55
Madhya Pradesh								
Mandsaur	Mandsaur	4	847	V-M,	22	142	116	51
Raisen	Gairatgunj	6	1181	V-M	23	82	91	41
Vidisha	Vidisha	4	1134	V-M, V-L	28	91	98	38

(Continued)

Annexure 5.1 continued

State/District	Taluka/Tehsil	No. of Villages	Rainfall[b] mm	Soil Type[c]	% of CPRs to Total Village Area	Persons/Km² of Total Area[d] in Taluka[d]	in the Village	Persons/10 ha of Area of CPRs in the Villages
Maharashtra								
Akola	Mangruipir, Murtizapur	5	840	V-M,	11	116	145	130
Aurangabad	Aurangabad	4	727	V-M,G	15	128	114	76
Sholapur	Akkalkot, Mohol, Sholapur (N)	4	667	V-D, V-M	19	122	111	59
Rajasthan								
Jalore	Sanchor, Bhinmal	5	421	S	18	67	98	54
Jodhpur	Jodhpur	3	319	S	16	68	81	50
Nagaur	Jayal	3	389	S, G	15	71	73	48
Tamil Nadu								
Coimbatore	Coimbatore, Pailadam	4	718	A-L, S	9	250	361	402
Dharmapuri	Dharmapuri, Pennagaram	3	844	A-L, S	12	286	210	169

[a]Based on district, taluka and village records and field work in the villages during 1982–5.
[b]Average annual rainfall of the nearest rain gauge stations of the study villages.
[c]Soil types: S = sandy and/or sandy loam, G = gravelly, A = alfisol (red soils), V = vertisol (black soils), D = deep, M = medium deep, L = shallow.
[d]Population and area data relate to 1981.
Source: Adapted from Jodha 1986a.

Annexure 5.2a. Management of CPRs: Measures Directed to the Protection of Area of CPR-Units in the Study Villages

Management Measures and Other Details	No. of Measures (events) for Different CPR Types						
	Comm. Pasture	Comm Forest	Wasteland	Watershed Drainage	Threshing/ Dumping Ground	Water Ponds, Catchment	Total
Complaint/protest against:							
Manipulation of land records	12	7	4	35	20	6	84
Enchroachment by people	17	14	15	32	24	12	114
Encroachment by government	7	10	4	8	8	7	44
Formal transfer of CPR area	9	5	7	11	12	9	53
Litigation against privatization of CPRs	8	6	-	10	6	6	36
Factional fights over privatization of CPRs	10	7	6	7	18	8	56
De-privatization of privatized CPR	6	4	-	12	15	4	41
Panchayat resolution for CPR-area protection	2	3	-	-	4	-	9
Nomination of hon. custodian watchman	8	4	3	3	13	6	37
Total	79	60	39	48	120	59	474

aIncludes river/rivulet banks.

Source: Case histories of 176 CPR units over 30 years ending 1982. Table adapted from Jodha 1990a.

Annexure 5.2b. Management of CPRs: Measures Directed to the Usage Regulation of CPR-Units in the Study Villages

Management Measures and Other Details	No. of Measures (events) for Different CPR Types						
	Comm. Pasture	Comm. Forest	Wasteland	Watershed Drainage	Threshing/ Dumping Ground	Water Ponds, Catchment	Total
Protest/action against:							
Cutting trees/bushes from CPRs	7	18	6	28	7	9	75
Removal of top soil from CPRs	9	6	4	12	-	-	31
Blocking access to CPRs by village influentials	14	12	6	27	22	6	87
Trespassing by outsiders, irregularity in product auction	18	21	3	20	-	-	62
Litigation/factional fight on misuse of CPRs	9	17	3	8	23	15	75
Village meeting to streamline auction procedure[a]	10	7	6	7	18	8	56
Agreement on periodical closure of CPR	6	4	12	15	4	41	
Provision of penalty for outside trespassers	17	16	12	-	-	45	

[a] Auction of fuel/dung collection rights, lopping of trees etc.
Source: Case histories of 176 CPR-units over 30 years ending 1982. Adapted from Jodha 1990a.

Annexure 5.2c. Management of CPRs: Measures Directed to the Development of Area of CPR-Units in the Study Villages

Management Measures and Other Details	No. of Measures (events) for Different CPR Types						
	Comm. Pasture	Comm. Forest	Wasteland	Watershed Drainage	Threshing/Dumping Ground	Water Ponds, Catchment	Total
CPR product auction, investment of revenue on CPR	19	20	4	3	12	16	74
Seeking government grant for CPRs	30	24	7	9	8	20	98
Contribution to CPR uplift							
Cash	4	3	-	-	-	9	16
Labour	22	12	7	17	10	14	82
Others	6	3	-	-	7	4	20
Physical work on CPRs							
Fencing	9	12	-	6	12	-	39
Trenching	12	7	3	9	-	-	31
Desilting/cleaning	-	-	-	11	-	12	23
Planting/protecting trees	6	15	10	7	14	8	60
Linking contribution to CPR uplift to other rituals/practices	6	9	-	-	10	13	38
Maintaining village bull by CPR-revenue	9	8	-	13	9	12	51

Source: Case histories of 176 CPR-units over 30 years ending 1982. Adapted from Jodha 1990a.

6

Grazing Lands and Biomass Management in Western Rajasthan
Micro-level Field Evidence

Rangelands or range resources in developing countries with high man–land ratio connote something entirely different from the concept as understood in developed countries. The differences primarily relate to the area or size of unit, ownership and access to resources, as well as expected type of output from these resources. These in turn influence the status, usage and upkeep of resources, and also the institutional arrangements governing them. From the users' perspective, the purpose of the range resources is to ensure the supply of total biomass, which besides fodder and forage, includes materials for fuel, fibre, fencing and even food. Hence, the assessment of current productivity and development requirements of range resources should extend to total biomass from them. Due to heavy pressure on land, areas formally declared as permanent pastures (range areas), often owned by the communities or government and accessible to the whole community, are too small to fully meet the total biomass needs of the people. Hence the supplies from rangelands are supplemented by biomass from other sources, such as crop lands and community forest. An important related issue is the prevalence of crop and livestock-based mixed farming systems in these areas. This reinforces the need and utility of the integrated use of diverse sources of biomass.

Thus, in developing countries range management isses are an integral part of the people's overall strategies for production and usage of biomass from different sources. The present chapter,

First published as 'Grazing Lands and Biomass Management in Western Rajasthan: Micro-level Field Evidence', *Annals of Arid Zone* (Special Issue on Range Lands) 34(2), 1995.

therefore, is focused on the wider issues of biomass production with usage, where not only fodder and forage have primacy, but other products and services of range resources also play an important role.

ARID REGION OF WESTERN RAJASTHAN: CONSTRAINED PRODUCTION ENVIRONMENT

The arid region of western Rajasthan, comprising eleven north-western districts, is characterized by highly erodable sandy and sandy loam soils with scattered rocky patches, low and highly erratic rainfall, deep and often saline ground water. On the basis of tehsil-level information on soil, vegetation and water resources, a team consisting of soil scientists, agronomists and conservation specialists, had broadly assessed the use capabilities of the land in the different districts (Anon 1960). They divided the region into three zones on the basis of land-use capability. Accordingly, the bulk of the region (zone I) consists of land classes VI and VII, suitable only for pasture and range development, according to the FAO use-capability classification of lands for conservation purposes. Less than a quarter of the region (zone II) contains land of classes VI and VII and tracts of land class IV. Class IV land could be put to restricted crop cultivation if accompanied by a number of conservation measures and practice of crop-fallow rotation. The remaining part (zone III) of the region contains lands of classes III (considered suitable for cultivation) and IV. After mapping the areas according to the classification indicated above Jodha and Vyas (1969) reported that nearly 79 per cent area of the region, unless transformed by irrigation, is not suited to high-use intensity invloved in crop farming (Table 6.1).

KEY CONSTRAINTS AND POTENTIAL

Despite intra-regional heterogeneities some important features of the region's natural resource base and their implications may be noted.

First, highly erodable and nutrient-poor soils and paucity of moisture do not permit intensive use of land, particularly in zones I and II. They favour activities like pasture-based livestock rearing as against crop farming. Secondly, low and unstable rainfall and short period of moisture availability cause greater uncertainty for production of grain (i.e. maturing of crop) as against producing other biomass from the same crops.

Thirdly, natural vegetation including desert trees, shrubs, bushes, perennial grasses (and even annual grasses at times) is relatively less sensitive to length of wet period and fluctuations in rainfall than domesticated crops. The former imparts greater certainty to production of biomass in the arid areas.

Historically, societal responses to these constraints (and potentials) have emphasized land-use practices and cropping systems which enhanced the quantity and certainty of biomass availability and thereby helped sustain livelihood activities without unduly increasing the use intensity of land.

SOCIETAL ADAPTATIONS

STRATEGY

Societal adaptations to arid environment may be seen in the form of various practices ranging from scattered settlement pattern (called *dhani*) and nomadism on the one hand to folk agronomy and ethno-engineering (involving a variety of measures designed) for conservation and security of water on the other. Nevertheless, our concern here being limited to highlighting societal responses *vis-à-vis* scarcity and instability of biomass supplies, we may look at the desert farmer's traditional systems designed to maximize production of biomass and combat its instabilities.

Chart 6.1 sketches the key components of the farmer's strategy. Indigenous agroforestry, crop-fallow rotation, CPRs, folk agronomy emphasizing biomass production, collective security measures against biomass scarcity, crop-livestock-based mixed farming, supply-led adjustments in the use of biomass, etc., constitute important provisions of the desert farmer's biomass-centred strategy. These provisions, individually or jointly, help in meeting his different objectives, including sustained supply of forage and fodder. This may be illustrated by highlighting the functions of key components of the strategy.

MICRO-LEVEL EVIDENCE

Mix of Extensive and Intensive Uses of Land: The farmer in arid areas tries to achieve the twin goals of higher biomass production, especially fodder, and lower use intensity of land (to avoid erosion) through a mix of several measures. The practice of crop-fallow rotation and provision of indigenous agroforestry are two important methods in this respect.

Crop-Fallow Rotation: The data from villages in Jodhpur and Nagaur

Chart 6.1 Objectives and Functions of the Farmer's Biomass-oriented Strategy in Western Rajasthan

Components of Strategy	Biomass Augmentation	Biomass Stability	Annual–Perennial Links	Extensive Land Use	Collective Security	Adjustment to Weather	Flexible Demand/Supply	Conversion of Biomass into Economic Gains
Indigenous agroforestry	✓	✓	✓	✓				
Common property resources	✓		✓	✓	✓			
Folk agronomy								
Crop-fallow rotation	✓			✓				
Cultivator's choice[a]	✓	✓		✓				
Crop livestock mixed farming				✓		✓	✓	
Supply-led management practices[b]						✓	✓	
Seasonal migration					✓	✓		✓

[a]Crop with high stalk grain ratios, high salvage potential.
[b]Extent of storage, processing, recycling according to relative scarcity/abundance of supplies.

Table 6.1. Indicators of Constrained Production Environment and Intra-regional Heterogeneities in the Arid Region of Western Rajasthan (11 Districts)

Features	Zone 1	Zone 2	Zone 3
Land use capability classes	VI, VII	VI, VII, IV	III, IV
Annual mean rainfall (mm)	< 300	300–500	500–700
Rainy days a year (no.)	13–19	20–27	25–35
Length of crop growing season (weeks)	9–11	12–16	15–20
Frequency of drought (crop failure) in ten years (no.)	5–7	3–5	2–3
Ground water : depth of wells (m)	120–150	75–100	15–50
Average size of:			
Landholding (1971) (ha)	19.7–24.6	9.9–14.7	3.5–6.4
Animal units per 10 persons (1977) (no.)	20–32	13–17	5–9
Farm activity favoured by land classes	Pasture-based livestock rearing, extensive type of land use	As under zone 1, and restricted cultivation with crop-fallow rotation	Normal cultivation and restricted cultivation on class IV land
Zone's share in region's:			
Area (%)	56.3	22.5	21.2
Population (1971) (%)	21.3	34.5	44.0
Livestock (1977) (%)	29.6	39.9	30.5
Cropped area (1980–1) (%)	28.3	37.6	24.1

Sources: Based on data culled out from report of the State Land Utilization Committee (Anon. 1960), District Census Hand Books (1971). Statistical Hand Books of different districts. Report of Agricultural Census 1971, Statistical Abstracts of Rajasthan (different years). All published by Government Press, Jaipur.

districts (belonging to zone II) indicated that 29 to 38 per cent area of the crop land was under fallow during 1963–5 (Table 6.2). This not only helped in rebuilding of soil fertility, but also provided space for grazing and collection of fodder, fuel and other non-crop material. Part of the land fallowed for a prolonged period, called *bira* (protected fallow plots), was done specifically for collection of fodder.

Provision of Indigenous Agroforestry: The number of trees, especially *khejri* and *ber* bush colonies called *malla*, within the plots and *matt* (shelterbelt) on the field borders, were different components of the indigenous agroforestry system in the study villages (Table 6.2, section B). Trees provided fodder, fuel and other material, besides keeping part of the crop land under natural vegetation. Khejri trees are lopped for fodder and fuel every year after the kharif (rainy) season harvest (Mann and Saxena 1980). Ber bushes with deep roots are cut (harvested) before planting the next crop. They resprout and grow with the crops without root competition. The fodder from khejri, called *loong*, and from ber bush called *pala*, are available every year. They are less affected by rain conditions than crop-based biomass. Hence they are called desert farmers' natural insurance measures (see Table 6.3, section C). During the drought years loong and pala fetch as high a price as the food grains. During good rain years the crop yields both grain and fodder, around khejri trees and ber bush colonies (malla) in the plots, are often higher than other parts of the same plots. Trees and bushes influence micro-climate and help the accumulation of humus. Unlike trees and ber bush colonies within the field, the trees, bushes and shrubs constituting matt are not harvested, except for fuel, thatching, etc., though camels and goats browse on them. Thus, agroforestry components perform several functions besides acting as renewable resources of fuel and fodder.

Folk Agronomy: Emphasis on high biomass production and extensive pattern of land use is also reflected by the choice of cultivars in crop farming (Table 6.2, section A). Crops like local pearl millet (*bajra*) and sorghum (*jowar*) with longer maturity period, indeterminate type (i.e. crops having recurrent flushes of flowering), and high stalk-grain ratios get higher preference. Similarly crops which have high salvage potential (i.e. possibility of harvesting fodder, if not grain, in the event of mid-season rain failure) get higher priority.

Complementary Use of Perennials and Annuals (Agroforestry): Provision of indigenous agroforestry clearly reflects this facet of the farmer's strategy for higher and stable production of biomass. Peren-

Table 6.2. Indicators of Extensive Pattern of Land Use, Emphasis on
Biomass Production, and Changes Therein at Farm Level in the
Villages Studied

Details	Jodhpur		Nagaur	
	1963–5	1982–4	1963–5	1982–4
A. Landholding level data				
Sample farms/holding (no.)	38	38	43	43
Average land holding (ha)	9.3	7.8	8.4	7.1
Proportion of land holding:				
Fallowed total[a] (%)	29	15	38	14
Fallowed as bajra				
(for fodder harvest; %)	11	4	30	9
Planted to crops	71	85	42	86
Proportions of cropped area				
devoted to crops with:				
Long maturity(%)	78	73	66	62
High stalk grain ratio (%)	62	65	74	73
High salvage value (%)	75	75	69	70
B. Plot Level data				
Total plots monitored (no.)	27	27	32	32
Khejri trees (no./ha)	20	26	18	29
Ber-bush colonies (no./ha)	8	3	15	9
	8	0	17	2
C. Average no. of above plots harvested				
during 10 years (1972–3 to 1982–3)				
For grain yield		6		5
For fodder from crops		8		7
For loong (topfeed trees)		10		10
For pala (topfeed shrubs)		10		10
Fencing/fuel material		10		10

[a]Unintended fallows due to rain failure are excluded.
Sources: Jodha 1968, 1986a. Two villages were selected in each district.

nials like khejri trees and ber bushes besides facilitating better performance of crop, act as more stable sources of biomass for the farmer.
Table 6.3 presents relevant details of biomass production from
perennials and annuals, including planted crops, in Jodhpur and
Nagaur districts.

Complementarity of CPRs and PPR: The emphasis on biomass
production and extensive use of land is not confined to household
and plot levels. At community level also there are several provisions
to ensure stable flow of biomass and encourage 'extensive' use of

land. These include keeping part of the land (often sub-marginal lands) as village CPR. CPRs are the resources which every member of the community can use. They include village pasture, forest, wasteland, watershed drainage, pond, etc. (Jodha 1985b). Villagers, particularly the rural poor, depend very heavily on the CPR. CPRs are an important source of forage, cut fodder, fuels and other materials. People supplement their own supplies of biomass by seasonal or year-round collection of material from CPR.

Table 6.4 presents some indicators of complementarity between CPRs and PPRs (private property resources) as a part of the desert people's strategy to ensure higher and stable supplies of fuel, fodder, and other material. Information is presented in the form of ratio between CPR and PPR to animal unit grazing days and ratio of supplies of fodder, fuel, other material from CPR and PPRs. The values are presented in Table 6.4.

If interpreted in terms of percentages the same data would reveal:

(a) Of the total land resources available to villagers, CPRs constituted 21 to 28 per cent in different districts during 1982–4. This figure was much higher (33 to 51 per cent) only thirty years ago (figures in parentheses in Table 6.4).

(b) During 1982–4, around 66 to 79 per cent of animal grazing was contributed by CPR. This was for all sample farmers and not only the rural poor as reported earlier (Jodha 1986a).

(c) About 55 to 65 per cent of fuel (including dung) collected by all sample households (including rural poor) was contributed by CPR.

(d) About 29 to 41 per cent of the collected fodder (non-crop products) was contributed by CPR.

(e) About 59 to 68 per cent of fencing and thatching material, which is ultimately recycled as fuel, was contributed by CPR.

The above facts clearly reveal the importance of CPRs in supplementing PPRs in meeting fodder and fuel needs at the village level in the arid region.

Collective Security against Periodic Scarcities: Rain-induced scarcity of biomass, especially fodder, is recurrent in the arid region. To soften its impact several institutional arrangements have been evolved by the desert people. The provision of CPRs, as discussed above, is most important among them. The concern for collective needs and mutual help is also reflected through what may be described as seasonal CPR. Accordingly, fellow villagers are allowed free access to private

Table 6.3. Average Estimated Biomass Harvest during Different Rainfall Years in the Villages Studied

Year	Rainfall (mm)	Fodder (kg ha⁻¹)			Food-grain (kg ha¹)	Fuel/fencing/ Thatching Material (cart load)	
		Topfeeds (long+pala)	Grass	Crop By-product		Crop By-product	Others
Jodhpur							
1963–4	159	377 (78)2	85	62	7	0	2
1964–5	377	305	780	2250	503	3	2
1965–6	295	318	250	435	108	2	2
1972–3	270	325	345	515	115	0	3
1973–4	473	342 (53)	1850	2160	615	4	2
1977–8	389	305	500	1250	212	0	3
1982–3	211	310	380	690	180	0	4
1983–4	378	246 (38)	320	1120	372	0	4
C.V. (%)	30	11	92	71	75	-	-
Nagaur							
1963–4	139	385 (80)	115	183	18	0	3
1964–5	472	380	1100	2500	612	4	3
1965–6	221	465	315	710	232	1	3
1972–3	198	420	400	915	214	0	4
1973–4	510	390 (48)	615	1240	558	3	–2
1977–8	521	405	930	2440	642	5	3
1982–3	256	440	640	915	312	0	–4
1983–4	410	335 (35)	938	1280	495	2	3
C.V. (%)	42	9	53	51	54	-	-

Note: Figures in parentheses indicate share of pala (ber bush fodder) in total quantity of top feed.

Data collected from the same farmers during different rounds of fieldwork revealed:

  a. During the low rainfall years top feeds (long and pala) play important compensatory role in overall supplies of biomass (i.e. fodder).
  b. During good rainfall years crop by-products contribute the largest share of total biomass produced.
  c. As seen in the field and indirectly revealed by Table 6.3 during good rainfall years, biomass from agroforestry components is not fully harnessed, partly due to labour shortage and partly due to sense of complacency. Not only pala and loong remain unharvested, but in many cases even bajra stalks are not harvested.
  d. During low rainfall years there is also greater emphasis on collecting material for fuel, fencing and thatching (including from CPRs, for own use, as well as for sale).

e. The last and most important inference suggested by the above table is the greater stability of biomass from perennials. This is revealed by the values of coefficient of variations reported in the last rows of the table.

Source: Based on field studies.

lands, for animal grazing during the post-harvest season, use of top feeds and collection of dung and other material. Similarly, under the system of periodic closure of parts of village territories to animals (called *chait rakhai*), individuals' rights to graze even on their own lands are suspended. The closure of territory coincides with the spring season, beginning around the month of Chaitra (late March–April), to permit regrowth and sprouting of perennials including trees (Chart 6.2). These practices are in decline. Contribution to charity feeding during scarcities is another indicator of collective risk-sharing system. Seasonal maintenance entrustment on mutually agreed terms (Jodha 1985b) is another measure to help each other. Most important, there are several informal institutional arrangements, including community sanctions, to regulate the operation of collective security measures. There is not enough quantitative information on these aspects except the data on their changes presented in Table 6.6.

Crop and Livestock-based Mixed Farming: Mixed farming based on crop production and animal rearing is one way to ensure balance between extensive and intensive uses of land as it does require keeping of some areas under natural pasture. Through linkages between farming, forestry, pasture and livestock, mixed farming helps in the diversification of sources of fuel and fodder and cycling of biomass.

Biomass Orientation of Folk Agronomy: Crop farming, despite its unsuitability in several areas, is an essential part of mixed farming systems in the arid region. Folk agronomy (i.e. traditional art and science of raising crops and managing resources) is not only sensitive to crop attributes like resistance to drought, etc., but it puts heavy emphasis on the fodder component of the crops. Accordingly, the desert farmer gives high priority to cultivars with high stalk-to-grain ratio, higher salvage potential, i.e. forage availability in the event of crop failure.

Supply-led Flexibilities in Biomass Management: In the arid areas of western Rajasthan, the supply of fuel and fodder or biomass, in general, fluctuates depending on the rainfall. The farmer changes his management (both harvest and usage) of biomass resources according to the quantum of biomass available. A comparative picture of

Table 6.4. Complementarity of CPRs and PPRs in Providing Biomass in the Study Villages, 1982–4

	Jodhpur	Nagaur	Jalor	Barmer
Ratio CPR:PPR lands[a]	2.4:7.6 (3.3:6.70	2.1:7.6 (4.8: 5.2)	2.2:7.8 (4.1:5.9)	2.8:7.2[b]
Ratio of contributions of CPRs and PPRs in animal unit grazing days	6.6:3.4	6.8:3.2	7.1:2.9	7.9:2.1
Ratio of CPRs and PPRs Contributions to collection/harvest of different items:				
Fuel	5.5:4.5	6.1:4.9	6.5:3.5	-
Fodder	3.5:6.5	2.9:7.1	4.1:5.9	
Fencing/thatching	6.2:3.8	5.9:4.1	6.8:3.2	-

Note: Data presented here relate to all sample households and not only the rural poor as reported in an earlier paper (Jodha 1986a).

[a]Figures in parentheses indicate situation during 1951–3 and are presented to show the decline in the area of CPRs.
[b]Data relate to 1963–5.

Source: Data and records collected for a study of CPRs by Jodha (1986a).

biomass production and consumption practices during the years with poor and good rainfall can illustrate the point. Table 6.5 presents data from the study villages of Jodhpur and Nagaur districts. During the low-rainfall years, the farmer undertakes several steps to augment production/collection of biomass (Table 6.5, section A). These include measures like collection of seeds as fodder or fuel to harvesting of premature crops. Most of these measures, being inferior options, are completely disregarded during the good-rainfall years.

Similarly, in the use of biomass, i.e. both fuel and fodder, a number of measures involving processing (e.g. chaff cutting) and recycling of biomass are adopted during the years of scarcity (Table 6.5, section B). The extent of these measures becomes negligible during the years with good rainfall.

Thus, the relative abundance of biomass and scarcity of labour during good rainfall years led to lower harvesting and consequent lesser conservation of biomass supplies. The opposite happens during years of scarcity. The difference in the farmers' approach to biomass management during good- and low-rainfall years has several implications. Most important is that non-harvested components of

Table 6.5. Varying Patterns of Management and Usage of Fodder and Fuel
during Years of Scarcity and Abundance in the Study Villages

| | Jodhpur | | Nagaur | |
	1963–4	1965–6	1972–3	1973–4
Rainfall (mm)	159	377	198	510
A. *Proportion of plots used for augmenting biomass supplies through* (%)[a]				
Collection of weed as fodder	33	3	48	0
Harvesting field borders	43	4	39	3
Harvesting grain crops as fodder	10	0	42	0
Premature harvesting of:	31	0	62	0
ber bush for fodder, etc.	58	0	70	2
khejri tree (lopping) for fodder and fuel	45	2	65	0
B. *Proportion of sample households[b] undetaking the following measures* (%)				
Used stalk				
after chaff cutting	98	14	100	36
without chaff cutting	2	86	0	64
Reused leftover fodder of productive animals for feeding unproductive animals	54	5	78	12
Mixed leftover foder (waste) for making dung cakes	60	6	94	20
Left bajra stalk, ber-bush unharvested	0	7	0	12
Collected fuel/fencing/thatching material from CPRs (including for sale)	48	16	52	20

[a]Number of plots monitored in the villages was 60 in Jodhpur and 88 in Nagaur district.
[b]Number of sample households covered the information was 62 in Jodhpur district and 75 in Nagaur district.
Sources: Field surveys for Jodha 1968, 1974, 1985b.

biomass (including sometimes crop stalk) during the good rainfall years represent a slack resource, which could be harnessed and through storage, used for meeting deficit during the low-rainfall years. The fact that non-harvest of top feeds (loong and pala) during a year adversely affects their yield in the succeeding year, adds to the importance of this slack resource.

The same applies to fodder (and even fuel) which is used without

Chart 6.2. Management of CPRs in Western Rajasthan: Whether Past Practices Continue Following Land Reforms

Practice	Whether Practice Continues
A. Indicators of private coast of use of CPRs	
Grazing tax (ghas man)	No
Fees for grazing in some CPRs on priority basis	No
Livestock-related levies (*laag baag*)	No
Compulsory labour contributions for desilting ponds (*begar*)	No
Penalties for disregarding grazing regulations[a]	No
B. Indicators of regulated use of CPRs	
Evenly scattered watering points	No
Deliberate rotation of grazing around different watering points	No
Periodical closure of parts of CPRs (e.g. *chairakhai*)	No
Periodical restriction on entry of animal category (e.g. sheep/ cattle) to parts of CPRs	No
Posting of watchman *(kanwaria)* with power to enforce regulation	No
C. Indicators of revenue earning[b]	
Auction of dung collection rights from CPRs	No
Auction of top feeds from CPRs	No
Auction/sale of wood from CPRs	Yes
Penalties for breaking grazing regulations	No
Cash and kind taxes and levies from users fo CPRs	No
D. Indicators of investment in CPRs	
Periodic desilting of ponds[c]	Yes
Payment to watchman *(kanwaria)*	No
Maintenance expenses of community bulls[d]	No
Support to scouts to survey water and fodder situation on migration routes during drought	No

[a]Panchayats also have provisions for imposing penalties, but such cases relate to trespassing by persons on migration routes during droughts, or to complain of damage to one's crops by others' animals, which are brought to panchayat officials for impounding.

[b]Feudal authorities collected substantial revenue from CPRs but reinvested only a small proportion of it.

[c]Periodic desilting of ponds now takes place through government relief expenses during drought years.

[d]Some Panchayats have provisions for maintenance of the bulls.

Source: Adapted from Jodha 1985b.

Table 6.6. Changes over Time in the Area of CPRs in the Arid Region of
Western Rajasthan (11 Districts) during 1951–2 to 1980–2[a]

Details	1951–2	1961–2	1971–2	1977–8	1980–1
CPR area (million ha)	11.3	9.8	9.2	87	8.4
CPR area as percentage of total geographic area	60.5	51.1	47.9	45.1	43.5
Percentage decline in CPR area over previous period	-	12.4	6.7	4.5	3.4

[a]Common property resources include forests, permanent pastures, cultivable and uncultivable wastelands and fallow lands other than current fallows. This table includes several items such as state forest (besides community forest), fallow lands and remote and inaccessible waste land, etc. This is because of non-availability of break up of data to give a precise extent of CPRs (as used in other tables). Hence, data presented here should be treated as broad indicators of status of CPRs.

Source: Adapted from Jodha 1985b.

processing (i.e. chaff cutting) during the years of abundance. A few closely monitored cases in the study villages showed that bajra and jowar stalks when fed after chaff cutting, can meet the need of two to three times more animals than is possible through unprocessed feeding of stalks.

PUBLIC INTERVENTION AND TRADITIONAL STRATEGIES UNDER STRAIN

PUBLIC POLICIES AND PROGRAMMES

In the changed circumstances, public policies and programmes, which represent formal institutional dimensions of resource management/ development, are being commented upon. Furthermore, only those public measures are discussed which directly influence the status and productivity of resources contributing to the supply of biomass. These measures relate to (a) land distribution policies; (b) usage regulation of land, especially the common grazing lands; and (c) land development and productivity promotion programmes (see Chapter 5).

Ever since the introduction of land reform programme in the 1950s, distribution of land by privatization of common property lands (used mostly for grazing) has been the major component of land policies in the state of Rajasthan and elsewhere in India. The policies were strongly welfare oriented (as indicated by 'land to the poor'

Table 6.7. Decline in Productivity of CPRs as Illustrated by the Histories of Four Forest and Grazing Plots in a Village of Nagaur District, 1964–5

Product	Plot 1 (6 ha)		Plot 2 (10 ha)		Plot 3 (12 ha)		Plot 4 (12 ha)	
	1945–7	1963–5	1945–7	1963–5	1945–7	1963–5	1945–7	1963–6
Timber (babul and indok trees)	12	3 (0)	11	1	3	0	17	0
Top feed (loong from khejri)	8	4 (2)	10	3	21	8	12 (5)	3
Top feed (pala from ber bush)	-	-	-	3	12	4 (0)	15	2
Fuel wood (khejri, ker, etc.)	8	2 (1)	5	2	18	6 (3)	21	4
Cut grass (kared and dhaman perennials)	13	3 (2)	18	4	27	9 (2)	21	0
Cut grass (bharoot, etc., annuals)	3	5 (5)	5	-	10	8 (5)	13	9
Dung Collection	-	- (1)	-	-	15	0	17	0
Gum (babul and indok trees)	40	0	10	-	-	-	-	-

Note: Gum is measured in kilograms. All other products are measured in cartloads. The weight of a cartload ranged from 500 to 100 kg depending upon the product (e.g., fuel wood versus top feeds) under question. By 1985 due to introduction of rubber-tyred bullock carts (Chhakada) the standard of cartload changed. Compared to earlier wooden-tyred bullock carts. Chhakada could accommodate 50 per cent more products by volume and weight. However, the figures reported in the Table are in terms of load carried by wooden-tyred bullock carts. Original sources of data are auction records of ex-Jagirdar and the village panchayat. In the post-land reforms period, the practice of auctioning has declined mainly because there is not enough material to auction. This in turn is a result of elimination of most of the trees and complete destruction of even roots of perennial grasses. Information originally collected for Jodha 1968.

Source: Adopted from Jodha 1985b. Figures in parentheses, added later, relate to 1982–4.

approach), and completely insensitive to use capabilities of land. This resulted in massive transfer of sub-marginal lands from natural vegetative cover to crop farming with low and uncertain productivity (Jodha 1985b).

The distribution of land was not accompanied by any obligation on the land recipients to use the land according to its use capabilities. Furthermore, there were no measures and provision to regulate the use intensity of land, both crop lands and grazing lands. The traditional informal arrangement regulating use of the land by rotational grazing, periodical resting of land, etc. got disrupted with the introduction of formal, legalistic system of village administration, represented by village panchayats.

For raising land productivity considerable efforts were made on the technological front. In the case of crop technology, however, the focus was on raising grain yield by high-yielding varieties, etc., with little attention to biomass used as fodder. The measures to raise productivity of rangelands, through a variety of methods like reseeding, use of chemical inputs, soil working, etc., were too much 'technique' dominated. They were quite insensitive to institutional factors, which condition the people's participation and adoption of technologies. Other policies focused on non-crop options on the land. The new initiatives in terms of agroforestry systems and silvipastoral programmes rarely crossed the boundary of research-cum-demonstration farms and pilot project areas, again due to the dominance of 'technique' and their crucial dependence on 'subsidy' to sustain them.

To sum up, the range-resource-related public policies and programmes in the arid zone, in the context of which farmers evolve their own response and strategy, were neither guided by biomass requirements, nor made sensitive to the use capabilities of arid land. Yet, as public interventions, they were strong enough to disturb the existing biomass-oriented farmers' strategy.

Apart from public policies and programmes, the rapid population growth and increased role of market forces have also played a significant role in the reduced feasibility and efficacy of traditional strategy. However, as Jodha (1985b) reported, public intervention has accentuated the role of the other two factors.

REDUCED EFFECTIVENESS

The Balance between Extensive and Intensive Uses of Land: This balance is disturbed by extending crops to sub-marginal areas and reduced extent of periodical fallowing of land. The proportion of land fallowed has declined from 29 to 15 per cent in Jodhpur villages and from 38 to 14 per cent in Nagaur villages during the last twenty years or so. Due to increased pressure of population, not many farmers can afford to leave land fallow (Table 6.2 section A).

Decline of Indigenous Agroforestry: As shown in Table 6.2, section B, several components of indigenous agroforestry, especially ber bush colonies (malla) and shelterbelts (matt) have declined significantly. Ber bush has disappeared due to recurrent use of tractors (mostly by hire) for cultivation (Jodha 1974). Tractor, unlike bullock-operated plough, cuts the deep roots of the ber bush. The roots, by regular regeneration, not only offered fodder and fuel but acted as soil binder. Disappearance of shelterbelts is largely due to encroachment by farmers, who gradually extend their plot areas beyond the legal field borders.

The only redeeming feature of the situation is that with the disappearance of ber bush, the farmers have increased their attention to protection of khejri trees in their crop fields (Table 6.2, section B). Thus, one source of biomass insurance (e.g. khejri for top feed) has partially substituted the other (i.e. ber bush for fodder). Decline of pala (ber bush fodder) production is also revealed by Table 6.3.

Decline of CPRs: Though CPRs perform several useful functions such as a source of collective sustenance during scarcity periods, a major source of fuel and fodder and a contributor to balance between intensive and extensive land uses, they have suffered the most both in terms of decline in their area (Table 6.6) and productivity (Table 6.7). Between the early 1950s and the early 1980s, the area of CPRs has declined by 37 to 63 per cent in the study villages of different districts.

In general, over-exploitation and under-investment have become the key attributes of the community's approach to CPRs. This has further exposed CPRs to degradation. Decline in their physical productivity is the final consequence. In the absence of any benchmark information, it is difficult to measure the decline in the productivity of CPRs. Yet, using records and oral history details, the situation is illustrated by Table 6.7. Accordingly, the yield of virtually all products of CPRs has declined.

To sum up, crop fallow rotation, indigenous agroforestry and CPRs are the key components of the traditional strategy to increase

Table 6.8. Indicators of Rising Scarcity of Fodder and Fuel in the Study Villages (Average Annual Situation)

Details	Jodhpur			Nagaur		
	1951–3	1963–6	1982–4	1951–3	1963–6	1982–4
Pachasa[a] (unit of fodder staking) in the village (no.)	21	13	0	35	18	3
Cart loads of fodder (stalks) contributed by the villagers to common/charity feeding (no.)	15	11	3	20	14	6
Proportion of households[a] sold fodder:						
On exchange basis (%)	28	22	8	44	28	14
For cash to trader (%)	6	30	43	16	36	62
Duration of seasonal outmigration of sheep herders (days/year)	28	42	98	20	60	112
Time it took for fuel gatherers to collect *mundana* (a cart full of fuelwood) for sale from village commons (days)	10	15	40	7	12	30
Proportion of households[b] stocked (traditionally discarded) inferior crop by-products for fodder and fuel (%)[c]	3	27	95	6	12	98
Proportion of farm households[b] allowed free access to others for post-harvest grazing, lopping trees and bushes, collection of dung, etc. (%)[c]	100	56	8	100	60	15
Households[b] replacing bush fencing:						
Every year	85	52	14	88	44	16
With a gap of 2–3 years	15	47	66	12	36	59
Replaced with stone fencing	-	1	20	-	10	25

[a] *Pachasa*, a form of staking fodder, secure for 5–8 years; lit, method of fodder storage for 50 years.
[b] Number of sample households was 62 and 75 respectively, from two villages each in the districts of Jodhpur and Nagaur.
[c] Include bajra husks, sesamum pods/stalks, mustard and raya stalks, and bengal gram stalk.

Source: Data collected during field studies for Jodha 1968, 1986a.

Table 6.9. Changes over Time in Livestock Farming in One Village Each in Jodhpur and Nagaur Districts, 1963–78

Farmers	Nagaur				Jodhpur			
	1963–65		1977–78		1963–65		1977–78	
	Small Farmers[a]	Large Farmers	Small Farmers[b]	Large Farmers	Small Farmers	Large Farmers	Small Farmers	Large Farmers
Average size of livestock holding (animal units)	15	13	13	9	16	14	15	9
Share of sheep/goats in (%)	38	6	42	22	40	9	46	31
Proportion of buffalo in milch stock (%)	5	23	13	46	6	27	15	51
Unproductive animals per productive animal (no)	7	4	6	2	5	3	5	1
Cattle regularly stall-fed (except in monsoon) (%)	6	25	11	49	5	23	18	57
Proportion of animal grazing days depending on CPRs (%)	59	81	59	76	31	85	62	76 29

Notes: Data relate to one village in each district. Details of the first four items relate to one village in each district. Details of only two farming groups are presented to indicate the contrast or comparison. The details of only two farming groups are presented to indicate the contrast or comparison, while the last two items relate to the village.
  [a] Those owning up to 5 hectares of dryland.
  [b] Those owning 10 or more hectares of dryland.

Source: Adapted from Jodha 1985b.

and stabilize the availability of biomass. But they have been adversely affected in recent years. The other provisions like collective security of biomass, supply-led flexibility in biomass management and crop-livestock-based mixed farming have been affected by changes in these three components.

INDICATORS OF INCREASING SCARCITY OF BIOMASS

Following the decline and degradation of resource base, biomass supply has declined in the arid region. Village- or farm-level data at three points of time (Table 6.8) indicate a drastic decline in the last three decades or so in the practice of fodder stocking (pachasa) at the village level and the villagers' contribution to common/charity feeding, to have been very drastic. Similarly, there is an increase in the duration of seasonal out-migration of sheep herders and in the time taken in the collection of a cartload of fuelwood (*mundana*), which indicate increasing biomass scarcity in the study villages. A shift towards increased use of cow dung in place of wood as a fuel was also observed in some study villages.

ADJUSTMENT TO INCREASING SCARCITY

Table 6.8 also presents some details which could be considered as indicators of both decline in the supply of biomass as well as adjustment to the declines. One such indicator is people's acceptance of inferior options. A number of biomass items like bajra husks (as fodder) and sesamum stalks (as fuel), which were traditionally discarded and allowed to rot are row stocked for use, even by rich farmers. Similarly, the tendency towards privatization of products of seasonal CPRs (i.e. crop fields during the off-season), reduced frequency of replacing old farm fencing out of material from khejri trees or ber bushes also reflects adjustment to reduced supply of biomass.

The shift towards private use of seasonal CPRs is partly an adjustment to decline in other CPRs and decline of collective arrangements against biomass scarcity. A prominent form of adjustment to biomass scarcity could be seen in the livestock sector of study villages. According to Table 6.9 (reproduced from Jodha 1985b) the livestock composition has undergone significant changes following the decline in CPRs as well as other changes like improved marketing of milk.

In the study villages not only has the average size of animal holding declined, the proportion of unproductive animals has also declined while the proportion of sheep and goat has increased. Unlike cattle they can not only manage on degraded pastures but have better facility to seasonally migrate to canal areas of neighbouring states of Punjab and Haryana. The increased proportion of buffalo can be attributed to milk marketing facilities under programmes like Operation Flood and better provision of drinking water in the village since the early 1970s (Jodha 1985b). Such milk animals seldom graze on CPRs, and in a way they represent withdrawal of rich farmers from CPRs. This sort of changes in the composition of livestock have been observed in many other dry tropical regions of India (Jodha 1986a).

FUTURE POSSIBILITIES

Better management of arid lands with focus on higher vegetative cover has been a key recommendation of several studies (Rai 1942; Anon. 1960; Jodha and Vyas 1969). The fragility of the region's natural resource base and its susceptibility to rapid erosion due to intensive use, its comparative advantage in pasture-based livestock farming, and the accentuation of desertification due to unscientific land-use practices were the key issues behind those recommendations. The specific suggestions ranged from formation of state-level board of land management to restriction of crop cultivation on sub-marginal lands and controlling the growth of animals. Some variants of the recommendations became part of public programmes like drought-prone area programmes (DPAP) and production-oriented relief strategy during droughts.

Researchers at the Central Arid Zone Research institute (CAZRI), Jodhpur, also provided the technical basis for public intervention like sand dune stabilization and improvement of rangelands. Despite these efforts, the situation of vegetative cover or biomass production and its proper usage has only worsened with time. The ineffectiveness of these initiatives may be attributed to their over-emphasis on 'techniques' (i.e. mechanical or biological dimensions) and insensitivity to institutional factors. These programmes betray a complete lack of understanding of the factors and processes at village and farm levels which are responsible for the rapid loss of vegetation or biomass in the arid region. The present paper, though covering only a small number of villages, has tried to put together micro-level evidence on the dynamics of resource use in the arid areas. The insights presented by the study, hopefully, would help reorient thinking on the problems and prospects of biomass-producing resources in the arid region. Among the realistic ways of promoting sustained availability of biomass in the arid areas of western Rajasthan is to sensitize the planning and development strategies in the region to the traditional or ongoing approaches of the farmer to biomass issues. This suggestion is not intended to idealize tradition but use it as a source of potential options, which could be improved with the help of advances in technology and management. Guided by this understanding, I propose to summarize the key inferences from the above discussion. The inferences are summarized separately for (a) measures dealing with augmentation of biomass availability, and (b) measures related to demand side of biomass. The discussion on each issue is presented in terms of current trends, possibilities of encouraging or discouraging the trends, and constraints on possible intervention to help the positive trends.

MEASURES AUGMENTING BIOMASS AVAILABILITY

REVIVAL OF INDIGENOUS AGROFORESTRY

Trends: There is an encouraging increase in the number of khejri trees in crop fields, but components like ber bush colonies (malla) and shelterbelts (matt) are rapidly disappearing.

Potential: Reasonably high chances exist for promoting trees in the field, as the farmer's current initiative is a response to a felt need for drought period insurance, especially after ber bush in the fields declined following tractorization. Decline of CPRs (i.e. their area and productivity) also encouraged protection and growth of trees (in crop lands) as a component of agroforestry to serve as an alternative source of biomass.

Public Intervention: At present there is little scope for effective public intervention, legal or otherwise. Some incentives to farmers might help. Discouragement of tractor cultivation may be necessary. The formal agroforestry programmes need to be sensitized to farmers' approaches to biomass production.

RETIRING CROP LANDS BACK TO NATURAL VEGETATION

Trends: This largely implies crop (grass/bush) fallow rotation. With rising population pressure decline in keeping the land fallow is unavoidable. Only large farmers are able to follow this practice.

Potential: There is very limited scope for this practice except on large farms, unless non-crop biomass production through new technologies becomes highly profitable.

Constraints: High pressure on land and tractor ploughing leave little chance for fallowing. Besides, there are difficulties in protecting the private gains like fuel, fodder, etc., from the fallow lands, since such fallows are also used as *de facto* common resources.

Public Intervention: There is little scope for any legal measures because of the inability to enforce any land-use regulation within the existing institutional framework. Moreover, no viable technologies for silvi-pastoral systems are available as yet. Hence there is a need for public effort in this direction.

REHABILITATION OF CPRs

Trends: CPRs continue to sustain people, particularly the rural poor, but the products available are increasingly inferior and less in quantity. Both the area and productivity of CPRs are in decline. Only limited evidence is available on the protection and rehabilitation of CPRs, where NGOs or enlightened panchayats took the initiative. In most cases the decision-makers, i.e. influential villagers, bother little about CPRs, as they do not depend on CPRs, except for grabbing them as private property.

Potential: There is little scope for improvement in the situation unless state policies and informal institutional arrangement at village level are changed or the NGOs take it up as a high priority activity.

Constraints: The key constraints are the state land policies which encourage privatization of CPRs, the panchayat's indifference to management and usage of CPRs which tends to turn CPRs into open access resources, and very low productivity of CPRs which also discourages any public initiative to manage and develop them.

Public Intervention: Public measures like afforestation, reseeding of rangelands, etc., are purely technical measures and are insensitive to the CPR dimension of these resources. This needs to be changed. The same applies to legal measures. The steps may include prevention of further privatization of CPRs, rehabilitation of CPRs with the help of NGOs (for usage regulation), and forest and soil conservation departments (for vegetative regrowth), incentives to panchayats to improve management of CPRs, provision of 'user cost' to serve both as a disincentive for over-exploitation and a source of revenue for upkeep of the resources. Finally the CPR dimension of all resource development programmes like social forestry, rangeland rehabilitation, etc., should be emphasized.

HARNESSING OF SLACK RESOURCES

Trends: Farm level evidence shows (a) non-harvesting of considerable biomass, and (b) insufficient conservation, processing and recycling of the biomass before use during the years of plenty. This represents a slack resource, which through processing and storage can help meet the needs during scarcity years.

Potential: At times these practices can add as much as 20 to 50 per cent to the overall availability of biomass, especially when one compares the quantity of sorghum/pearl millet stalk fed to animals with and without chaff cutting. Similarly, during any year, the harvest of top feed loong from khejri tree and pala from ber bush declines considerably if they are not harvested in the previous year.

Constraints: The relative scarcity of labour during the good-rainfall years and the sense of complacency generated by plenty, lead to disregard of this biomass potential; slackening of the traditional practice of on-farm reserves and recycling of biomass are the other problems.

Public Intervention: Incentives through fodder prices, fodder bank, popularization of fodder/fuel processing and recycling methods using modern technologies may help, though as yet there is no formal initiative on this front.

INTER-REGIONAL COMPLEMENTARITIES

Trends: The system of seasonal migration of animals, both as a form of collective security and method of adjustment to instability of biomass availability is rapidly changing, mainly because of the reduced area of grazing CPRs. Migration of cattle is in decline, while migration of sheep and goat has become an annual feature. The latter migrates on a regular basis, to 'green revolution areas' of the neighbouring states, after the rabi harvest. They make good use of grazing material, a waste for high-biomass-producing areas, and also help in sustaining the organic base of 'green revolution', which otherwise is largely dependent on chemical fertilizers.

Potential: Having emerged in response to felt needs, the above pattern is self-sustaining. This is an indirect way of making up the deficit of biomass in the arid areas and deficit of organic manure in green revolution tracts. Some people transport the animals by trucks. It is likely to increase in future.

Constraints: The usual problems which animal migrants often face during transhumance, e.g. harassment *en route*, etc., are key constraints.

Public Intervention: Facilities for migrants, e.g. fodder and water facilities *en route* and protection against harassment especially on state borders, can help the sheep herders.

DISCARDING ZERO-INPUT PRODUCTION SYSTEMS

Trends: Biomass production (other than crop by-products) in the arid areas involves little cost except on labour for harvesting or grazing the biomass. The people simply harness what nature offers them. A few large farmers who tried sheep penning and a little scratching/trenching of fallow lands got almost double the yield of biomass, as compared to the fallow lands without any development input.

Potential: In view of the tradition of harnessing nature at zero cost and inability to protect the gains of investment in resource improvement for oneself (due to the convention of common access to private fallow lands), the system of producing fodder and fuel with delibrate investment and effort is almost non-existent. The situation may change with the availability of highly productive silvi-pastoral systems.

Constraints: Absence of tradition and practice to treat fodder/fuel as a crop among the people and low priority to them in agricultural R and D are the key problems.

Public Intervention: Introduction of modern silvi-pastoral systems is one potential option. Past efforts on this front, however, have been negligible compared to extension and support systems for arable farming. Moreover, R and D efforts on silvi-pastoral systems are completely insensitive to farm-level realities.

MEASURES ON THE DEMAND SIDE

On the demand side of biomass, the issues could be listed under two categories: (a) those relating to regulated use and prevention of over-exploitation of resources producing fuel and fodder, and (b) those relating to reduction in pressure of demand on biomass resources, both by raising the use-efficiency of biomass (e.g. processing preceding their use) and by reducing the overall demand on biomass (e.g. by gradual reduction in the number of animals).

REGULATED USE AND PREVENTION OF OVER-EXPLOITATION

Trends: Despite low biomass production potential, the vegetative growth of plants in several parts of the arid region is reasonably fast, if the plants are given sufficient protection and are utilized within the limits (e.g. through rotational grazing or through cut-and-carry system rather than open grazing). As already mentioned some investment in terms of manuring (e.g. by sheep) and moisture conservation measures can raise biomass productivity of land quite substantially. The resources to which the above description applies are CPRs and private fallow lands expected to be under natural vegetation. For private fallow lands, especially large farms, the trend is towards greater protection of resources.

Potential: As mentioned earlier, with the rapid decline of CPRs, the private sources of fuel and fodder are gaining importance. Private forestry (in bira, i.e. old protected fallow lands) as against social forestry on CPRs, shows greater potential for spread. However, this would make the rural poor worse off, as they have little land resources to participate in this process.

Constraints: Constraints to regulated use of CPR have already been mentioned. On private fallow lands, absence of community sanctions, enabling the protection of private gains of investment, etc., obstructs the process. Decreasing access to biomass resources for the rural poor may also indirectly obstruct the rapid rise of private forestry.

Public Intervention: Present usage status of CPRs offers a vast potential for public intervention. Introduction of usage regulations, incentives to panchayats and NGOs to implement them, introduction of grazing policies to help rotational grazing and periodic closure of CPRs are a few of the possible areas for public intervention. However, this will need initiative from below, involving NGOs and village communities. Some physical measures such as spread of watering points (as in the past) in grazing lands can also help. Introduction of usage charges for CPR users is another (fiscal) measure to induce regulated use of CPRs and generate revenue for the upkeep of CPRs. Farm- or village-level storage, conservation and recycling of biomass can also be promoted by certain fiscal measures.

REDUCING PRESSURE OF DEMAND

Trends: There are two major ways of reducing pressure of demand on biomass, especially the fodder resources: (a) conservation and processing of biomass before use, and (b) reduction in the number of animals. Regarding the former, despite the high potential of slack resources, the trends indicate increasing de-emphasis on storage, conservation and recycling measures. Regarding the latter, at least in some areas there is a clear trend towards reduction in the number and changes in the composition of livestock. Livestock farming is becoming more management-intensive, involving high private cost. This, in the long run, may encourage decline in herd size. However, this trend forms a part of the emerging pattern, where large farmers are increasingly depending on private resources for their livestock and the rural poor have to sustain their animals on rapidly shrinking and degrading CPRs.

Potential: The need for the utilization of slack resources and discarding of surplus livestock cannot be overstated. However, the potential for the adoption of these practices seems quite low. An integrated approach directed at demonstrating economies of small herds can help. Increase in the extent of stall feeding and consequent reduction in herd size in the areas with improved milk marketing facilities (e.g. 'operation flood areas') in western Rajasthan are possible new directions.

Constraints: The old tradition of raising animals at social cost and the dependence on free supplies of fuel and fodder from CPR are the major obstacles to reduction in demand for biomass. Security through large number of low-productivity animals is another constraint to reduction in animal numbers. These factors apply more to the rural poor, who do not have enough private resources to complement their use of CPRs.

Public Intervention: Infrastructure, marketing, formal insurance, etc., can help promote stall feeding based on high management-intensive livestock rearing, especially for cattle. 'User group'-based development and management strategies for CPRs can be one way to help the rural poor in regulating their number of animals.

Present livestock policies and programmes are not sensitive to these issues.

7

Biophysical and Social Stresses on Common Property Resources*

1. INTRODUCTION

Rural common property resources, also known as common property regimes/ resources (CPRs) are institutional arrangements evolved by communities to collectively manage and use their natural resources. In India's dry tropical areas (and in other dry regions) they also formed a part of rural people's strategies for adjusting to the harsh and stressful environmental conditions (Berkes 1989; Bromley and Cernea 1989). Historically speaking, in such areas CPRs may be treated as a product of the stressed environment. In the changing circumstances in India's dry tropics, however, despite the persistence of biophysical stresses, CPRs are rapidly declining. This has happened due to several changes accompanying the process of rural transformation, which have marginalized the role and utility of CPRs in the rural economy and have eroded the social and institutional framework that ensured protection and regulated use of CPRs as community assets. Thus, the man-made circumstances disfavouring CPRs have acquired primacy over the biophysical conditions that favoured CPRs. In the broader context, this may amount to disregard of ecological and environmental imperatives by focusing on short-term considerations. The present chapter based on a detailed study of CPRs in India's dry tropics illustrates this phenomenon.

The essay draws on the household- and village-level data collected during field work over three years from over eighty villages from nearly twenty districts scattered in six dry tropical states (see Map 5.1). Methodological and other details are reported elsewhere (see

*First published as 'Common Property Resources and the Environmental Context: Role of Biophysical Versus Social Stresses', *Economic and Political Weekly* 30(51), 1995.

Chapter 5). This essay analyses the field-level information with focus on the changing nature of stresses and the consequent changes in the status of CPRs in India's dry regions.

In the following section, a generalized picture of environmental features of dry regions that historically favoured the institution of CPRs is presented. This understanding is reinforced by a comparison of the situation of CPRs in areas with higher and lower degree of environmental stresses. The third section briefly comments on the persistent utility of CPRs and people's dependence on them, despite which there has been rapid decline in the area, productivity and upkeep of CPRs. In the fourth section, decline of CPRs is explained in terms of new sources of stress emerging from the general dynamics of rural change, demographic pressure, side-effects of public intervention/market forces, technological changes, etc. These factors seem to influence the community's approach to their natural resources more than the biophysical stresses that necessitated stronger CPR support for the village economies.

The conclusion of the chapter is that unless the biophysical constraints in the dry areas are substantially reduced, the deliberate marginalization of CPRs would mean reduced range of locally managed and used options for the people to withstand the environmental stresses. This is more so for the rural poor who continue to significantly depend on CPRs for their sustenance.

2. CIRCUMSTANCES FAVOURING CPRs IN INDIA'S DRY REGIONS

CPRs could be simply described as the community's natural resources where every member has access and usage facility with specified obligation, without anybody having exclusive property right over them. In the Indian villages, CPRs generally include community pastures, community forests, wastelands, common dumping and threshing grounds, watershed drainage and village ponds and rivers and rivulets with their banks and beds. Even when the formal legal ownership of some of these resources rests with certain agencies (e.g. wastelands or uncultivated lands belong to the state's revenue department) *de facto* they belong to the village communities. Most of these community resources, despite their other specific uses serve as important sources of biomass (fodder, fuel, food, fencing, timber, etc.) for the rural communities (Jodha 1990a). Equally important are the temporal and spatial patterns of supplies of their product, that further enhance the utility of CPRs in dry regions. To understand these and related aspects, we may comment on the agro-climatic features of dry tropical areas that impart special significance to the provision of CPRs in these habitats. Chart 7.1 sketches the key factors involved.

The production environment of India's dry-land areas is broadly characterized by low and variable rainfall, frequent droughts, heterogeneous (including erodable and low-fertility) land resources, nature's low regenerative capacities and limited and high-risk production options. These factors or constraints have several implications at the regional, village-community and farm-household level. The persistence of such harsh and risky environmental conditions gave rise to circumstances favourable to the provision of CPRs. For instance at the regional level (i.e. macro-units of dry land and tracts), the low and unstable production possibilities restricted population growth, encouraged market-wise isolation of villages, and did not attract enough technological and institutional intervention. All these circumstances offered limited incentives and compulsions for privatization of vast land areas. This helped in retaining fragile lands as CPRs. At the village-community level, the heterogeneity and fragility of land resources along with the variable rainfall made it difficult to fully harness the potential of resources and adequately meet the environmental risks through private-resource-based crop farming alone. Balancing of intensive (by cropping) and extensive (by pasture/forest) uses of land, as required by the resource characteristics, became a part of collective strategy for risk management and production enhancement. The provision of CPRs, enforced through the social sanctions for protection and usage regulation, facilitated the aforesaid strategy. At the farm-household level, despite several folk-agronomic practices such as crop- and livestock-based mixed farming, diversified cropping, and other elements of land-extensive farming systems, the narrow (farm) production base of private farming could not ensure protection against risks due to temporal and spatial variability of rainfall. Hence dependence on collective risk sharing and complementarity of PPR (private property resource) and CPR-based activities became necessary. This again favoured the provision of CPRs.

The features of agro-climatic environment and adaptive measures described above may be observed in smaller or larger measure in most parts of the dry regions. But the picture often gets blurred when the situation for macro-units (even district level) is aggregated. In the

Chart 7.1. Circumstances Historically Associated with CPRs

Natural Resource Base and Agro-Ecological Features

(Low and variable precipitation; heterogeneous including submarginal fragile land resources unsuited to intensive use; nature's low regeneration capacities; limited and high-risk production options, etc.)

Implications and Imperatives at:

Regional Level	Community Level	Farm Household Level
a. Low population pressure; market isolation; limited technological and institutional interventions.	a. Heterogeneity, fragility of resouce base; inadequacy of private risk strategies	a. Narrow, unstable production base; diversified, biomass centred, land extensive farming systems
b. Limited incentives and compulsions for privatization of CPRs	b. Balancing extensive-intensive land uses; focus on collective risk sharing	b. Reliance on collective measures against seasonality and risk
c. Overall circumstances (a, b) favourable to CPRs	c. Community responses to (a, b): CPRs (protection, access usage, etc.)	c. Induced by (a, b) stronger focus on complementarity of CPR–PPR (private property resources)-based activities

Source: The author.

process the micro-level (e.g. village-level) environmental situation which ultimately influences the status of CPRs remains unobserved (Chambers 1990). Hence, in order to illustrate the situation summarized by Chart 7.1, two subsets of villages out of the eighty-two villages covered by the study (Jodha 1986a), have been made, group one consisting of twenty-eight villages having higher degree of environmental stress and group two, of twenty-two villages with lower degree of biophysical stress, which influence the extent of CPRs and complementarities between the CPR-PPR based activities.

Some variables indicating environmental stress such as rainfall and its variability may not reflect the true picture at a particular village level, as these data often relate to rainfall at faraway locations, e.g. district or block headquarters, and the spatial variability of rainfall in dry tropics (even within a short distance of a few kilometres) is incredibly high (Virmani et al. 1982). Hence, while selecting the villages with high or low environmental stresses higher weight has been given to the indicators of the situation that is visible, observed or recorded at the village level. They include frequency of drought and crop failures, length of crop-growing season, extent of sub-marginal land, extent of land without irrigation facility, etc.

Summarized quantitative details (Table 7.1), from the villages with higher and lower degree of environmental stress may further corroborate the above inferences on CPR-promoting circumstances. According to Table 7.1 the villages with lower but highly variable rainfall, higher frequency of drought, shorter crop-growing season, and larger extent of sub-marginal lands, limited irrigation facility—all of them representing different sources of environmental stress—have greater extent of CPRs. Prior to land reforms of the 1950s (which substantially reduced their area), CPRs accounted for 39–58 per cent of total area in the villages with higher degree of environmental stress. The corresponding figures for environmentally stable (low-stress) villages were 15 to 23 per cent. CPRs, as mentioned earlier, represented part of the collective risk-sharing and resource-management systems, whose overall extent was much higher in the villages with greater biophysical stresses.

Not only CPR area and CPR contribution to production systems differ significantly between the villages with higher and lower degree of biophysical stresses, the extent of other adaptation measures, especially those focusing on the importance of biomass stability and its uses, also varied in the two groups of villages. This is reflected by

Table 7.1. Extent of CPRs and Other Collective Risk Sharing Strategies in Villages with High and Low Levels of Environmental Stress

Stresses and Strategies	Situation (range of values of the variables) in the Villages with:	
	High Environ- mental Stress (villages 28)	Low Environ- mental Stress (villages 22)
Indicators of stress		
Annual average rainfall (mm)	300–740	800–1150
Rainfall variability (coefficient of variation)[a]	33–39	18–21
Length of crop growing season (days)	65–90	185–220
Events of drought/crop failure in 5 year (no)	2–3	0–1
Area of sub-marginal lands in village area[b] (%)	69–82	8–13
Extent of irrigated crop lands (%)	0–6	10–33
Adaptation measures		
Households with dominance of livestock in mixed farming (%)	68–84	4–9
Households with natural vegetation as principal source of (fodder) biomass (%)	38–52	5–7
Proportion of area under crops with high stalk-grain ratio (%)	71–93	27–38
Extent of collective, sharing practices in the village [c] (no)	9–13	3–5
Households using more than four CPR products as input in private farming (%)	76–84	13–27
Share of CPRs in village areas 1950–2 (%)	39–58	15–23
Population density 1951(no/km^2)	37–49	105–182

[a]Coefficient of variation of rainfall based on rainfall records at district/taluka headquarters.

[b]Sub-marginal lands include areas with sandy and unfertile soils, high extent of salinity, rocky and undulating topography, area suffering with water logging, perennial weeds, shrubs, etc. not suitable for cultivation.

[c]Collective sharing activities include collective upkeep and protection of CPRs, common use of private lands during non-crop season, seed sharing, desilting of village ponds, maintenance of catchments of percolation tanks, joint field operation during crop season, fodder stocking for charity, maintenance of village bulls, contributory fund for common facilities (including joint litigation for village interests), etc.

Source: Data collected for the study of CPRs, Jodha 1986a. It covered 82 villages from 6 states in dry tropical regions of India. The distribution of sub-sets of villages, i.e. those with higher and lower degree of biophysical stresses respectively, is as follows. Andhra Pradesh (3, 4), Gujarat (4, 5) Karnataka (4, 3), Madhya Pradesh (4, 2), Maharashtra (4, 3), Rajasthan (6, 2), and Tamil Nadu (3, 3).

higher values of relevant variables in the drier villages. For example the extent of natural and produced biomass-centred strategies in the drier villages was higher, as reflected by values of different variables (such as crops with higher stalk-grain ratio, dominance of livestock in mixed farming, CPR products as major inputs in private farming, etc.). Table 7.1 provides more comparative details on the two sets of villages studied (see also Chapter 5).

3. DECLINE OF CPRs DESPITE THEIR RATIONALE AND UTILITY

Information under Chart 7.1 and Table 7.1 indicated that CPRs played am important role in people's biomass-centred and diversified production systems and represented an important component of collective resource-management and risk-sharing strategy in India's dry tropical regions. This may be complemented by some quantified information on the contributions of CPRs.

Despite the monitoring and measurement complexities, the contributions of CPRs were quantified particularly in terms of fuel and fodder supplies as well as employment and income generation for rural households. Details for over eighty villages reported elsewhere (Jodha 1986a) indicated that rural people, especially the rural poor with very limited private resources, depend on CPRs for meeting the bulk of their biomass needs. Over 80 to 100 per cent of the poor households depend on CPRs for these supplies. According to information summarized in Table 7.2, despite degradation and reduced productivity, CPRs in different areas contribute to the poor households as follows: fuel supplies 66–84 per cent, animal grazing 69–84 per cent, employment 128–196 days per household per year, annual income Rs 534–774 per household. This constitutes more than a fifth of total household income in most cases. Rich households' dependence on CPRs (for products, income, employment) is very little. Their focus is on acquiring CPR land as private land. Inclusion of CPR income in household incomes reduces the extent of rural income inequalities, as indicated by Gini-coefficient of income distribution (Table 7.2).

Table 7.2. Extent of People's Dependence on CPRs in India's Dry Regions (CPRs Contribution to Households Supplies, Employment, Income, etc.)

State (with No. of Districts and Villages)	Household Category[a]	Fuel Supply[b] (%)	Animal Grazing[c] (%)	Per Household Employment[d] Days (no.)	Per Household Annual Income[e] (Rs)	CPR Income as Proportion[f] (%)	Value of Gini-coefficient of Income from[h] All Sources	All Sources Excluding CPRs (%)
Andhra Pradesh (1, 2)	Poor	84	-	139	534	17	0.41	0.50
	Others	13	-	35	62	1	0.41	0.50
Gujarat (2,4)	Poor	66	82	196	774	18	0.33	0.45
	Others	8	14	80	185	1	0.33	0.45
Karnataka (1,2)	Poor	-	83	185	649	20	3	
	Others	-	29	34	170			
Madhya Pradesh (2,4)	Poor	74	79	183	733	22	0.34	0.44
	Others	32	34	52	386	2	0.34	0.44
Maharastra (3,6)	Poor	75	69	128	557	14	0.4	0.48
	Others	12	27	43	177	1	0.40	0.48
Rajasthan (2,4)	Poor	71	84	165	770	23	-	-
	Others	23	38	61	413	2		
Tamil Nadu (1,2)	Poor	-	-	137	738	22	-	-
	Others	-	-	31	164	2		

[a] Number of sample household from each village varied from 20 to 36 in different districts. 'Poor are defined to include agricultural labourers and small farm (<2 ha. dryland equivalent) households. 'Others' include large farm households only.
[b] Fuel gathered from CPRs as proportion of total fuel used during three seasons covering the whole year.
[c] Animal unit grazing days on CPRs as proportion of total animal unit grazing days.

ᵈTotal employment through CPR product collection.

ᵉIncome mainly through CPR product collection. The estimation procedure underestimated the actual income derived from CPRs (see Jodha 1986a).

ᶠCPR income as percentage of income from all other sources.

ᵍHigher value of Gini coefficient indicates higher degree of income inequalities. Calculations are based on income data for 1983–4 from a panel households covered under ICRISAT's village level studies (Walker and Ryan 1990). The panel of 40 households from each village included 10 households from each of the categories, namely large, medium and small farm households and labour households.

Source: This and all other tables in this chapter are based on village/household data from study villages reported by Jodha 1986a.

However, despite environmental imperatives supporting the need for CPRs and quantifiable evidence on their contributions to the rural economy (especially the economy of the rural poor), since the early 1950s, CPRs are in decline in every part of India's dry tropical regions.

Table 7.3, reproduced from an earlier report (Jodha 1990a), covering all the eighty-two villages from seven states in the dry regions indicates that CPR area has declined by 31 to 55 per cent in the study villages of different states during the early 1950s to the early 1980s. Other studies in the dry regions referred to in Chapter 5, have also identified the accentuation of this decline.

First, according to Chart 7.2 at the regional level, due to rapid population growth and the state's undeclared policy of converting CPRs into private lands, and technological and market-related factors, there are more opportunities and stronger incentives (or compulsions) to convert CPRs into private lands.

At the overall regional level, on the basis of evidence from over eighty villages, the role of these factors has been discussed elsewhere (see Chapter 5). In the sub-grouped villages, where higher degree of biophysical stresses suggests the need for extensive CPRs, the role of factors diversely affecting CPRs is illustrated in Table 7.5.

Accordingly, a comparison of the situation during the early 1950s and early 1980s shows that population density in most villages has almost doubled; distance to market centres has reduced substantially; technological change in terms of irrigation facility and tractorization has increased significantly; land prices have increased threefold; feasibility and opportunity for privatization of CPRs, which were non-existent, have become real and attractive possibilities (as shown by the extent and events of privatization). All these components of rural transformation have created new incentives or compulsions to reduce CPR area.

At the village-community level, these developments are complemented by the decline of group action or collective strategy for resource management and risk sharing. This happened both due to increased differentiation of the rural community as well as marginalization of traditional forms of rural cooperation. Usurpation of the community's mandate and initiatives by the state through a variety of legal, administrative and fiscal measures (Jodha 1990b) and the invisible role of market forces in changing people's attitudes towards collective measures have significantly contributed to these changes. It is seen from Table 7.5 that the number of collective measures or

Table 7.3. Extent and Decline of Area of CPR Land in India's Dry Regions

State (and no. of districts)	No. of Study Villages	Area of CPRs 1982–4 (ha)	CPRs as Proportion of Total Village Area		Decline in the Area of CPRs since 1950–2 (%)	Persons per 10 ha. of CPR Area	
			1982–4 (%)	1950–2 (%)		1951 (no.)	1982 (no.)
Andhra Pradesh (3)	10	827	11	18	42	48	134
Gujarat (3)	15	589	11	19	44	82	238
Karnataka (4)	12	1165	12	20	40	46	117
Madhya Pradesh (3)	14	1435	24	41	41	14	47
Maharastra (3)	13	918	15	22	31	40	88
Rajasthan (3)	11	1849	16	36	55	13	50
Tamil Nadu (2)	7	412	10	21	50	101	286

Note: CPRs include community pasture, village forest, waste land, watershed drainage, river and rivulet banks and other common lands. Data indicate average area per village.

Source: Adapted from Jodha 1986a, where more disaggregated details are reported.

Chart 7.2. Current Circumstances Adversely Affecting the Extent and Status of CPRs

Recent Economic Institutional Technological Changes Influencing the Patterns of Resource Use

Increased physical and market integration, increased extent and changed nature of public interventions, increased demographic pressure, etc. shaping the pace and pattern of rural development

Implications and Imperatives at

Regional Level:	Community Level:	Farm Household: Level
a. Population growth accentuating land hunger	a. Development led differentiation of rural community and decline of collective strategies for resource management, risk sharing etc.	a. Reduced area and productivity of CPRs, marginalising their contribution to diversified and biomass-centred production strategies
b. Public polices enhancing legal/illegal opportunites for CPR privatisation	b. Usruption of community's mandates, initiatives by the state through legal, administrative and fiscal means	b. Individualization of adjustment measures against risk, seasonality etc.
c. Technologies and market forces activating the land market, extending to fragile lands	c. Emphasis on acquiring CPRs as private property, rather than use collectively	c. Reliance on private resource, public relief, non-biomass oriented technologies, etc.
d. Over all circumstances (a, b, c) unfavourable to CPRs	d. Due to (a, b, c) rapid erosion of community concerns and group action for CPRs	d. Due to (a, b, c) reduced reliance on complementary of CPR-PPR (private property resources) activities/products

Source: The author.

group action activities (including management of CPRs) have declined from 9–13 during (or prior to) 1950–2 to 4–5 during the early 1980s. The state that hardly acted at the community level prior to the 1950s now performs 6–8 activities (which were performed by the community itself in the past). This is a consequence of increased interventions by the state under welfare and development programmes (Jodha 1996b).

A related consequence of the above is the increased importance of 'individually operated' as against the 'collectively operated' measures in farming systems.

Reduced area of CPRs causing overcrowding and over-exploitation of their potential has led to physical degradation and reduced productivity of CPRs. The over-exploitation and poor upkeep of CPRs, accentuating their physical degradation, is also due to slackening or discontinuation of traditional CPR management practices. As reported elsewhere (see Chapter 5), more than 90 per cent of the villages currently do not enforce usage regulation nor collect any levy or tax for investment in CPRs as they did in the past. More than 80 per cent of the villages no longer enforce any user obligations today. This is a consequence of the state interventions marginalizing the role of community leadership and decline of collective concern at the village level.

Table 7.4 presents the net results of the foregoing in terms of details on various aspects of physical depletion of CPRs. Findings by other researchers (see Chapter 5) also corroborate this evidence. Reduced number of products from the CPRs (from 27–46 in the past to 8–22 at present) and decline in the quality and quantity of existing products clearly manifest the biophysical decline of CPRs and their reduced capacity to sustain the biomass-centred economy of the villages in dry regions. This also represents the reduced biodiversity maintained through CPRs in the past (see Chapter 8).

PROCESS OF DECLINE OF CPRs: NEW SOURCES OF STRESS

The fact of rapid decline of CPRs in dry areas is well recognized. The factors contributing to the decline are also documented at macro- and micro-levels. The purpose of this section is to identify the process by which the role (rather than the presence) of biophysical factors supporting CPRs is marginalized. In other words, we look at the

Table 7.4. Some Indicators of Physical Degradation of CPRs[a]

Indicators of Changed Status and Context for Comparison	States (with no. of villages)						
	Andhra Pradesh (3)	Gujarat (4)	Karnataka (2)	Madhya Pradesh (3)	Maharashtra (3)	Rajasthan (4)	Tamil Nadu (2)
No. of CPR products collected by villagers[b]							
In the past	32	35	40	46	30	27	29
At present	9	11	19	22	10	13	8
No. of trees and shrubs per hectare in							
Protected CPRs[c]	476	684	662	882	454	517	398
Unprotected CPRs	195	103	202	215	77	96	83
No. of watering points (ponds) in grazing CPRs							
In the past	17	29	20	16	9	48	14
At present	4	13	4	3	4	11	3
No. of CPR plots where rich vegetation, (indicated by their nomenclature), is no longer available	-	12		3	6	4	15 -
CPR area used for cattle grazing in the past, currently grazed mainly by sheep/goat (ha)[d]	48	112	95	-	52	175	64

[a]Based on observation and physical verification of current status (during 1982–4) and the past details collected from oral and recorded description of CPRs in different villages (Jodha 1986a). The choice of CPRs where plot-based data are reported was guided by availability of past information about them.

[b]Includes different types of fruits, flowers, leaves, roots, timber, fuel, fodder, etc. in the villages. 'Past' indicates the period preceding the 1950s and 'Present' indicates the early 1980s.

[c]Protected CPRs were the areas (called 'oran', etc.), where for religious reasons live trees and shrubs are not cut. The situation of CPR plots (numbering between 2 to 4 in different areas) was compared with other bordering plots of CPRs which were not protected by any religious or other sanctions.

[d]Relates to area covered by specific plots, traditionally used for grazing high productivity animals (e.g. cattle in milk, working bullocks or horses of feudal landlords). Due to their depletion, such animals are no more grazed there.

Table 7.5. Quantified Details on Some Changes Adversely Affecting the Extent and Status of CPRs in the Study Villages[a]

Details of Change	Range of Values of Change During	
	1950–2	1982–4
Population density (no/km^2)	37–49	69–98
Distance from nearest market centre (km)	18–26	7–21
Cropped area cultivated by tractor (%)	0–1	18–69
Cropped are irrigated (%)	0–6	3–18
Cost of dry lands at 1980 prices (Rs./ha)[b]	450–700	1500–2500
Extent of CPR area privatised (%)	0–0	30–63[c]
Incidents of CPR-privatization		
Land distribution camps by government (no.)	0–0	8–12[c]
Illegal land grabbing cases (regularized) (no.)	0–0	18–26[c]
Community-level activities[d] done by		
Villagers'group action (no.)	9–13	3–5
Government agencies (no.)	0–3	6–8
During drought/scarcity households mainly depending on[e]		
Public relief (no.)	5–9	73–82
CPR products, collective supplies group action (no.)	63–80	15–17
Households using (> 4) CPR products as farm input (no.)	76–84	18–22
Proportion of CPR area in total land of the village (%)	39–58	16–28

[a]The data relate to 28 villages with high degree of biophysical stresses (Table 7.1).
[b]Based on limited number of land transaction in different villages.
[c]Data indicate the cumulative situation since the land reforms during 1950–2, rather than during 1982–4 only.
[d]For type of community level activities see note (c) under Table 7.1.
[e]Information relates to early 1960s and late 1970s for over 15 villages of Rajasthan, Gujarat and Maharashtra for which studies on impacts and adjustments to drought were conducted (Jodha 1978).

new sources of stresses disfavouring CPRs and their primacy over the CPR-supporting environmental stresses. For doing so, we first sketch the process in Chart 7.2 and then present some quantitative evidence to support it in Table 7.5. Furthermore, we present the quantitative evidence with reference to the subset of twenty-eight villages

(Table 7.1) where the higher degree of biophysical stresses necessitates the strong provision of CPRs.

Broadly speaking, the key factor adversely affecting the status of CPRs is the overall pattern of rural transformation, which has either reduced the importance of CPRs in the rural economy or made it difficult to maintain them as dependable community assets. The process is manifested by several factors such as increased extent of technological and institutional interventions by the state, physical and market integration of dry areas, increased population pressure and significant changes in people's attitude towards common resources, and visible changes .in farming systems and resource-use practices induced by new technological and market circumstances as well as state support. These factors, as elaborated below, have individually or jointly generated incentives or compulsions to discard CPRs at regional, community and farm-household levels. In the process, circumstances that historically favoured provision of CPRs at all the three levels (Chart 7.1) have been replaced by those which disfavour CPRs (Chart 7.2). These changes are felt and observed in most villages in the dry regions. Some of them have been reported in quantitative terms as they relate to the study villages (Table 7.5).

At the farm household level, due to their reduced productivity and output, CPRs have become less dependable components of farmers' biomass-centred, diversified production strategy. Furthermore, again due to facilities of public relief, development aid, new technologies and associated subsidies, etc., the individual risk-management measures have replaced the collective ones including the CPR-PPR complementarities.

The above description of the situation is corroborated by the quantified evidence presented in Table 7.5. Accordingly, if the proportion of involved households is any indicator, the dependence on CPRs during droughts and scarcity has reduced to around one-fourth of the past extent of dependence. The corresponding dependence on public relief and grants has increased manifold. Similarly, in place of 76–84 per cent of the households in the early 1950s, only 18–22 per cent had used CPR products as farm inputs in 1982–4.

However, it may be noted that within the rural communities, the rural poor still depend significantly on CPRs. But, like the rural rich, they too are active in acquiring them as private property. As a combined consequence of factors at regional, community and household levels, the proportion of CPR area in village lands has declined from

39–58 per cent in 1950–2 to 16–28 per cent in 1980–2 in different villages. Furthermore, more than 90 per cent of the privatized CPR lands, despite their fragility and unsuitability for cropping are put under crops with very low crop yields (See Chapter 5).

Thus the state's undeclared assault on CPRs; specific opportunities created by market forces; land hunger accentuated by population growth; collapse of traditional forms of rural cooperation; and reorientation of farming systems de-emphasizing the role of biomass, are the key factors that have led to the marginalization of CPRs' role and decline of their area and productivity in the dry areas.

Viewed in a broader and long-term context of sustainable use of dry lands, the changing situation of CPRs represents a well-known conflict between ecological and environmental imperatives of resource characteristics and the features of development interventions. At the moment both policy-makers and village communities seem to give higher priority to the latter. Consequently, CPRs are deliberately ignored as a part of development strategies (see Chapter 5). The present patterns of rural transformation do not support them. However, the ecological imperatives (supporting CPRs) and sustenance of rural poor are quite important factors, which may not be ignored unless substitute options to CPRs that meet these two concerns are evolved.

III. Ecosystem Revival

8

Biodiversity Management in Agricultural Landscapes[*]

 Quite probably, the total area of plant biodiversity outside protected areas is higher than in protected areas. The two broad land-use categories potentially contributing to biodiversity management are cultivated lands, particularly highly diversified cropping areas in developing countries; and uncultivated lands, particularly community lands or common property resources (CPRs) including community forest, pasture, watershed drainage, etc. The farmer's fields qualify as a habitat for *in-situ* conservation and management of biodiversity, due to their following features, observed particularly in areas not affected by monocropping-dominated agriculture.

(a) Such cultivated lands are largely planted to local land races of diverse crops and help in their utilization and propagation, especially in relatively inaccessible and non-modernized agricultural zone.

(b) Their use is dominated by folk agronomic practices involving diverse combinations of crops, resource-regenerative practices including crop rotations, organic recycling, complementary uses of annuals and perennials, etc., all of which are conducive to conservation, and maintenance of biodiversity.

(c) Acting as a base for integrated farming systems, these lands facilitate effective and mutually reinforcing linkages between different components favouring *in-situ* biodiversity management through harnessing of annual–perennial complementarities; crop-livestock complementarities, farming forestry linkages, etc.

(d) They are well-recognized sources of diverse land races of major cultivars as any germplasm collection mission in the past would

*First presented as 'Social Dimensions of Biodiversity Management in Agricultural Landscapes', in a meeting on 'Mainstreaming Biodiversity in Agricultural Development', Environment Department, the World Bank, 1996.

support. Crop breeding continues to depend on breeders' access to these land races; and the latter's maintenance by the farmer is an important contribution to biodiversity conservation.

(e) Such lands are generally dominated by small holdings, where diversification, use of local (often on-farm produced) inputs, resource regeneration and recycling constitute important components of farmers' adaptation strategy against risk and insecurity. Such strategy contributes to *in-situ* maintenance and utilization of biodiversity. The farmer's courtyard or home garden is another well-recognized area for promoting and using multiple plants and species, including medicinal herbs.

(f) 'Uncleared and crowded cropping plots', characterizing farming systems is another feature of such lands, especially in the case of small holders. This description implies: uncleared field borders habitated by shrubs, grasses and not currently used plants; only selective removal of 'weed' in the field; bush fences acting as natural shelter for wind-carried and bird-carried germplasm and its propagation as well as periodical use, especially during scarcity; and lack of thorough clearing and preparation of soil comparable with commercialized sole-cropping systems, where every plant except the chosen crop or species is treated and removed as a weed.

In keeping with these features of crop-lands we use the term 'biodiversity in backyard' while referring to them. They effectively complement the uncultivated, common lands as habitats of local-level biodiversity (Box 8.1).

Recognition of non-cultivated areas (including community forests, pastures, watershed drainages, sacred groves, community dumping grounds, common shelterbelts, wastelands, etc.), clubbed under common property resources (CPRs) as habitats for *in-situ* management and utilization of biodiversity by people is much easier. Unless CPRs are devolved into open access resources and degraded by unregulated use, they are under natural vegetation and depending on the soil-climate conditions serve as a rich source of biodiversity at local or community level (see Chart 8.1). Besides meeting people's multiple needs, they act as support lands for crop lands by offering organic input (e.g through farming–forestry linkages), regulate micro-level nutrient and moisture flows and (wherever relevant) act as buffer zones for protected areas. People have formal or informal institutional

BOX 8.1. THE WORLDWIDE INTEGRATION OF FARMLAND AND WILD-
LAND RESOURCES

Indonesia: Complex home gardens in Java have been found to contain
hundreds of species within a single village (Soemarwoto and Conway
1991). A range of annual and perennial crops are grown together
complementing the main rice crop. There are several different types of
gardens: intensively managed home gardens, the village/forest gardens
and the forest fringe gardens. The importance of wild foods increases in
gardens toward the forest fringe (Michon 1983).

The Philippines: Home gardens are also important for experimentation
with new varieties derived from wild species. The Hununoo traditionally
considered over Swidden 1,500 plants to be useful and cultivated about
430 of these in their fields (Conklin 1954).

Botswana: The use of a diversity of species is not limited to forested
areas. The agro-pastoral Tswana use 126 plant species and 100 animal
species as sources of food (Grivetti 1979).

Brazil: A study of the agro-forestry systems of a Brazilian family living
in the Amazon estuary shows how they harvest various native and exotic
species from a house garden, and managed flood-plain forest and
unmanaged flood-plain forest and within the managed area, some vines,
shrubs and trees are cut. The unmanaged area contains economically
important species such as the ace palm (*Euterpe oleracea*) and rubber
(*Hevea brasiliensis*). Together these three zones provide fish, game, fruits,
medicines, household items and oilseeds, for home consumption and
for sale. With the cash, the family is then able to buy other staple goods
(Anderson et al. 1985).

United Kingdom: In the Middle Ages manorial estates appear to have
been highly sustainable systems. This sustainability was not achieved as
a result of high productivity but because of the integrated nature of
farming and the great diversity of produce, including wild resources.
Wild resources were important for food, fodder for livestock, green
manuring and various household goods. They were carefully managed
at the local level through by-laws that varied from village to village. The
expansion of agriculture into common property lands led to increased
gross agricultural production but declining woodlands, pasture and
marshland resources and critically a loss of buffers for the rural poor.
This contributed to agricultural recession and the eventual decline of
the manorial system (Pretty 1990).

Note: References as cited in the source document.
Source: IIED 1995.

Chart 8.1. Important Interrelated Features of Community Management of Biodiversity

Feature	Implications/Imperatives vis-à-vis Recent Changes
Local management biodiversity (BD) is largely usage driven (i.e. its presence and promotion directly linked to meeting people's production/consumption needs).	Non-usability, distortions or change in use of biodiversity as input or output (through external substitute) weaken the community management of bio-diversity; 'usage' of BD needs to be integral part of biodiversity conservation strategies.
Biodiversity is seen more as an integrated bundle of resource and services to meet multiple needs rather than a sector or product-centred phenomenon.	Disintegration or breakdown of functional linkages of different components of biodiversity (as resources or services) through segregated sectorial or product centred interventions, which weakens the local management of biodiversity; BD promotion strategies should start with conceiving the integrated picture of services/products it could offer to local people.
Biodiversity management especially the set of institutional practices is strongly linked to site-specific local knowledge; autonomy and access to resources; and participatory practices.	Imposition of external arrangements insensitive to local needs and circumstances weakens the local management of biodiversity; use of traditional knowledge systems and user participation should be integral parts of BD promotion strategies..
Biodiversity management is strongly context specific, i.e. it is effective under diversified, and extensive type or resource use systems linked to low pressure and diversity of demand.	Recent changes, due to population growth and pressure of market demand, reduce the scope for people-managed biodiversity unless measures to promote diversification with intensification are evolved; incentives and pay-offs are ensured for biodiversity-friendly, diversified, low intensity, often low productivity activities (e.g. through organic certification and premium on organic products).

Source: The author. Also see Chart 8.4 for specific practices reflecting people's concern and action for biodiversity conservation.

arrangements to protect, develop and use CPRs and in the process promote usage-driven management of biodiversity.

SOCIAL DIMENSIONS OF BIODIVERSITY MANAGEMENT

The recent attention to social dimensions of biodiversity conservation recognizes the centrality of people and their approach to biodiversity resources. This also represents a stage in the evolution of formal approaches to biodiversity conservation promoted by the international community and national governments. Gradual recognition of the role of local communities in biodiversity conservation is manifested by successive stages in biodiversity conservation policies and programmes. Accordingly, the initial focus on 'protected areas' was subsequently supplemented by attention to buffer zone; this in turn at a still later stage has been supplemented by emerging focus on rural commons and finally on the private crop lands in non-modernized areas, as biodiversity habitats at local levels.

Strengthening of the biodiversity conservation (management) strategies at lower (micro) levels may contribute to strengthening of biodiversity management at successive higher (macro) levels, both by reducing the pressure on them and by promoting the culture of conservation through the bottom-up approach it involves (Box 8.2). The need for restructuring biodiversity policies/plans by incorporating human and social dimensions and designing approaches to implement the same, are now increasingly emphasized by both social scientists and natural scientists dealing with the subject (Miller 1995).

In keeping with the thematic focus of our discussion, the social aspects elaborated below are confined to the last two areas of biodivers-

---

Box 8.2. Genetic Diversity for Future Agriculture

Much of the genetic diversity on which the improvement and future sustainability of agriculture must depend is found in and around farmers' fields, in village woodlands and in grazing lands.

In-situ management of wild genetic resources is likely to be the most effective conservation method in the long term.

Incorporation of indigenous crops and other native plant germplasm in the design of self-sustaining agro-ecosystems should ensure the maintenance of local genetic diversity available to farmers.

Source: IIED 1995.

ity management, i.e. commons and crop lands, comprising agricultural landscapes and the people's involvement therein. In a way, the discussion is focused on a few central issues such as: do people (i.e. local communities) have any understanding and concern for biodiversity conservation? If yes, then how do they respond to them? Despite people's stakes in local biodiversity management, why and how is rapid erosion of biodiversity taking place in agricultural landscapes? What are the possible approaches to arrest and reverse this erosion process? And finally, how to integrate these approaches into biodiversity conservation strategy at different levels?

PEOPLE'S PERCEPTIONS AND PRACTICES

While commenting on the rural people's perceptions of biodiversity and its management, it should be recognized that for a variety of reasons, the perceptions could be captured largely through understanding of the resource management practices followed by them rather than by formally recording of their views on the subject. The resource-use practices (and the institutional and technological arrangements to support them), based on the physical proximity and understanding of local resources, evolved through a process of trial and error over generations, in a way represent the codification of people's perceptions and concerns relating to biodiversity. Even when they have acquired the status of a routine or a ritual, the traditional resource management practices have usable scientific rationale behind them. Hence, closer understanding of biodiversity management practices can help capture people's perspectives on the subject (Gupta 1991).

A brief account of biodiversity-related practices followed by farmers in different areas covering both crop-lands and CPRs will illustrate the point. The biodiversity management practices followed by farmers on crop-land have already been indicated.

In the case of CPRs, protection of area and vegetation, and their regulated use through various social sanctions or group action, etc. constitute the key tasks. The extent and nature of the practices in terms of the basis and method of resource sharing, dispute resolution, penalty for violators of rules, periodic investment for their upkeep (without external assistance), etc., differ from region to region. However, despite variation in the nature and extent of biodiversity management practices followed in different areas, they do have broad similarities in their orientation and dominant features.

DOMINANT FEATURES OF PEOPLE-MANAGED BIODIVERSITY

A closer look at the people's biodiversity management practices recorded by different studies reveals their dominant features. These features may not only reflect on why and how people conserve biodiversity, but can offer important elements for incorporation into

---

BOX 8.3. SOCIAL CONDITIONS AND CONSERVATION STRATEGIES

Social pressure is a major reason why natural forest management schemes are abandoned or disrupted mainly because the managed forest areas are invaded by local poor people. This was clearly seen in Colombia, where the Cartono de Colombia forest management project was disrupted when unemployed people proceeded to extract poles and to mine in the forest, thus destroying the regeneration capacity of the management circles. Similarly, community involvement, or sharing in the benefits, has often been one of the critical aspects leading to SNFM success.

The special need of local communities must be taken into account when designing forest management schemes so that they have an incentive to protect the forest and maintain the scheme. Where local communities hold traditional claim to forest land, secure land rights are a prerequisite to ensure that benefits from forest management flow to them. This was demonstrated in the ejidos of Quintana Roo, Mexico and in Palazcu, Peru. In both cases, job security and community economic stability improved as a result of the projects. The BOSCOSA project in Costa Rica is likely to lead to the same results. On the other hand, the Carton de Colombia example shows what can happen when local people do not benefit from the management programme. Thus, the private concessionaire, as well as the public sector, must consider local needs and develop mechanisms to channel revenues accordingly. While efforts to involve local communities in forest management schemes may initially appear difficult and time consuming, in fact, protection by local people usually costs less than government protection, and is believed to be more efficient.

Local people generally have knowledge of the forests and of the non-timber forest products that can be incorporated into management schemes. In fact, traditional forest management is generally criticized for not taking into account other forest products. Where possible, more reliance on extraction of fruits and gums and less on timber should help to reduce the environmental impact of extraction activities and create incentives for sustaining the natural capital. A good example of this harvesting chicle and honey, along with mahogany, in Quintana Roo, Mexico.

Source: Kirmse et al. 1993 as cited by IIED 1995.

national biodiversity conservation strategy and also ensure cost reduction as well as people's participation in the process (Box 8.3).

As summarized under Chart 8.1, people-managed biodiversity is: (a) highly use-driven in the sense that they protect and promote bio-diversity because they use it: and for the above reason; (b) biodiversity is conceived as an integrated bundle of services and products to meet their multiple needs rather than a sectoral or product-centred pheno-menon; (c) local resource knowledge (especially on the part of women), autonomy and access to resources significantly influences the status and management of biodiversity by the people as shown by a contrast between relatively inaccessible, externally less impacted areas and the mainstream urban-impacted, commercialized agricultural areas; (d) management of biodiversity conservation is strongly context-specific, accordingly, its feasibility and efficacy are very much linked to factors such as lower pressure on resources and diversified demand for products, which in turn promote diversified and low-intensity use of land resources conducive to biodiversity maintenance; (e) because of the above features, local biodiversity management practices and measures tend to acquire the status of routine and ritual, and become invisible to the mainstream decision-makers. This in turn influences the public policies and programmes in terms of their indifference and unfavourable orientation toward biodiversity management by the local communities.

THE CHANGING STATUS OF PEOPLE'S BIODIVERSITY MANAGEMENT SYSTEMS

Undoubtedly, people-managed biodiversity management systems are rapidly weakening. Chart 8.1 summarizes the implications and impera-tives of dominant features of people-managed biodiversity. They also reflect on the incompatibility between some features of biodiversity management systems, and the present-day changed circumstances. For instance: (a) reduced dependence on or usability of local biodivers-ity; (b) breakdown of integrated farming system (and resource-use systems) affecting usable biodiversity as a bundle of services and products; (c) marginalization of local communities and knowledge systems by interventions from above; (d) changed demographic, institutional and technological context undermining the feasibility of farmers' practices (e.g. low-intensity and diversification-oriented resource use, etc.) that promoted biodiversity maintenance; (e) policy-

Chart 8.2. Creating Space for People's Biodiversity Management Systems (PBDMS) in the Mainstream Work

Project the Relevance and Usability of PBDMS for Mainstream Work	Develop the Approaches and Methods to Document and Utilize PBDMS
(a) For general BD conservation: Contribution of BD in 'bush' and 'backyard' to global stock of BD Usage driven focus of PBDMS as a case of win-win situation and its other features as input for national strategies (b) For agriculture: Entry point for action on participatory, community-based initiatives in agricultural R&D and natural resource management BD in 'backyard' and in 'bush' as source of usable germplasm for R&D People's (specially women's) knowledge and experience with agro-diversity as untapped resource for R&D	(a) Information generation on: Status of PBDMS Its changes over time Factors causing change Ways to strengthen PBDMS Using PRA/SA[a] method; and relevant records/statistics (b) Identification of good practices: Using above methods and experiences of new initiatives by NGOs and others (c) Awareness generation and advocacy, focus on targeted agencies: Policy-makers, donors, R&D planners, NGOs Target group specific dissemination approaches (d) Action research and application: Involve people, R&D groups, BD and agricultural planners Incentive systems for people (e.g. premium on organic products, etc.)

[a]Participatory Rural Approval/Social Assessment.
Source: The author.

makers' persistent interventions, etc., are some of the changes which despite their possible other gains have adversely affected the people's biodiversity management, both in 'bush' and 'backyard'. Chart 8.3 provides an indicative list of public interventions adversely affecting biodiversity 'bush' and 'backyard'. Table 8.1 provides quantified information on the impact of land reforms in India on biodiversity status through decline of CPRs. In the light of these changes and unlikelihood of reversal of the above in the near future, advocacy of people-managed biodiversity systems may appear a futile exercise (Brandon 1995).

BOX 8.4. PROTECTION OF SEEDS BY PEOPLE

Though slowly declining, a few practices followed by the people in rural areas of India to protect and regenerate biodiversity are worth reporting:

(a) The practice of planting all available types of crops (whether fully used or not by the farmer) to maintain diversity of crops offered by the nature is still in vogue in many areas. For instance, in parts of UP hill region practice of planting *barahanaja* (lit. twelve types of seeds) is still followed. *Navdhanam* (lit. planting nine crops) in dry Telengana region of Andhra Pradesh.

(b) To restore and guard against the disappearing plant species the people in different villages follow practices as indicated below:

- Scattering in the fields the soil collected from below the thorn or bush fence, where germplasm accumulates and remain shelterled (and even regenerates and spreads) without any disturbance for years or even decades.
- Scattering of accumulated fine dust containing germplasm which settles overtime in the feeding structures meant for stall feeding of animals.
- There are sacred groves in villages (called Orsan in some areas) where cutting of trees or shrubs or even grass is ritually prohibited. People periodically sweep these areas to seek the favour of deities. The swept material (soil containing germplasm of different plants) is scattered around the crop fields and grazing lands. This helps in regenerating plant species already lost from the field due to overuse.
- Similarly sweeping of spots and scattering of swept material in the fields is periodically done in the case of places where unthreshed crops or fodder reserves are stocked or crop threshing is regularly done.
- In dry areas like Rajasthan, people hang waterpots on trees for

(Continued)

Box 8.4 continued

birds especially during the non-crop season. The soil and bird-dropping (containing seeds of different plants) are collected and scattered in the fields/pastures.

- In some villages the soil from around ants colonies are also collected and scattered in the fields and grazing land (believed to kill poisonous plants?)
- In frequently droughts affected villages, people carry seeds with them during outmigration and back. Purity and security of seed are said to be the main consideration behind this practice.
- In relatively cohesive and animal husbandry-dominated villages, the practice of protection of parts of village pasture on rotation basis against grazing until after formation and maturity of seed (to ensure regeneration in the successive years) is still popular.
- Finally, in several areas (including ecologically better endowed areas), there is a concern and effort toward biodiversity conservation. These efforts are based not only on concerned NGO efforts but the people's rising consciousness of seed as symbol of security, freedom and self-reliance. Such sentiments were visible in several areas (especially UP hills, Karnataka and Gujarat villages). The controversies relating to Dunkel proposal on intellectual property rights as part of GATT discussion and efforts of the farmer's lobby have also contributed to this. NGOs dealing with environmental and indigenous knowledge systems have helped in enhancing the consciousness on biodiversity conservation.

Source: Jodha 1995b.

Nevertheless, the following may be stated. Though there are very limited areas where one can see people-managed systems in their totality, their individual components are still practised in wider areas. In several cases, out of necessity their modified forms have emerged, representing: (i) compromise between (a) intensification and diversification; (b) productivity growth and conservation; or (ii) responses to external pressures/incentives and internal needs for resource protection. Furthermore, the hope for people-managed system is sustained by continuation and revival of traditional practices (Boxes 8.4, 8.5, 8.6) as well as a few emerging new possibilities (Kothari 1997). The latter are reflected by: (a) rapidly accumulating evidence on greater effectiveness and lower costs of participatory management of natural resources; (b) experiences of limited and scattered but quite impressive success stories of revival of people-managed

Chart 8.3. Indicative Dimensions of Public Policies/Programmes Influencing Local Resource Users' Actions Affecting Biodiversity

Indicative (interrelated) components of public policies and programmes	Aspects of local level biodiversity represented through changing situation of crop-lands (CL) and CPRs					
	Habitat context (landscape/ land-use changes)		Species context (product types and their extraction patterns)		Management context (local control, knowledge and need-based options)	
	CL	CPR	CL	CPR	CL	CPR
Policies/programmes with unfavourable side effects on the people's biodiversity management systems*						
Land Policies						
Land tilling, distribution, curtailment of CPR area (privatization) for transfer to crop farming		✓				
Agricultural Intensification						
Narrow focus on limited crops/ attributes disregarding diversification and reduced need of input and services from local biodiversity, regenerative/integrated farming systems				✓		✓
Agricultural R&D						
External input focussed product-centred rather than resource-centred R&D and technologies; focus on limited food crops rather than food system supported by CPRs and diversified cropping; focus on homogenization of cropping and agronomic practices	✓		✓		✓	

Pricing and Trading Focus
External linkages and marketing focused on extracting niche, disregarding diversity/minor crops

Centralized, Top-down Impositions of Generalized Interventions
Administrative, legal, fiscal and technical measures disregarding: local knowledge, autonomy and participation, folk agronomy, and regenerative practices

Biased Fiscal and Infrastructural Systems
Taxes, subsidies, incentives, foreign aid, support systems, etc. directed to reduced diversification, supporting limited crops and land-use types

Note: Policies and programmes with orientation opposite to the following can also be put in the same format of a bi-variate table.
Source: The author.

Table 8.1. Some Indicators of Loss of 'Biodiversity in Bush' (CPRs) Following the Land Reforms Intervention in India[a]

Indicators of Changed Status and Context for Comparison	States (with number of villages)						
	Andhra Pradesh (3)	Gujarat (4)	Karnataka (2)	Madhya Pradesh (4)	Maharashtra (3)	Rajasthan (4)	Tamil Nadu (2)
CPR-Products collected by villagers:[b]							
• In the past (no.)	32	35	40	46	30	27	29
• At present (no.)	9	11	19	22	10	13	8
Per hectare number of trees and shrubs in:							
• Protected CPRs[c]	476	684	662	882	454	517	398
• Unprotected CPRs	195	103	202	215	77	96	83
Number of watering points (ponds and protected catchments) in CPRs (also promoting BD):							
• In the past	17	29	20	16	9	48	14
• At present	4	13	4	3	4	11	3
CPRs (as BD habitats) as proportion of total village area[d]:							
• In the past (%)	18	19	20	41	22	36	21
• At present (%)	11	11	12	24	15	16	10

[a]Table adapted from Jodha 1992 (Also see Box 8.4).
[b]Includes different types of fruits, flowers, leaves, roots, timber, fuel, fodder, etc. in the villages. 'Past' indicates the period preceding the 1950s. 'Present' indicates the early 1980s.
[c]Protected CPRs were the sacred groves (called 'oran') where for religious reasons live trees and shrubs are not cut and grazing is restricted.
[d]This information relates to many more villages than the ones indicated in row one. Total number of villages covered was 82 from 21 districts of the above 7 states.

Box 8.5. Save the Seed Movement in UP Himalayas

This initiative acquired visibility and farmers' attention during 1990–1, really started as a result of a social worker-cum-farmer's concern towards emerging crop crisis in Hemval valley region of Tehri Garhwal (Himalayas) in UP. In this mountain valley where introduced high-yielding varieties of rice and white soyabean (supported by agricultural research and support systems) had made a considerable headway, the drought and pest attack of 1987–8 came as a major shock. While these crops completely failed, the local varieties of crops in remote areas of the region did quite well. Impressed by this, a social worker, Mr Vijay Jardhari of Jardhargaon in the valley collected seed of 15 local varieties of rice from remote areas (where modern varieties were yet to replace the local land races), and planted them on his field. During the next season while introduced rice varieties faced several pest damage the local varieties were not affected by pest. This convinced the local farmers about the superiority of local varieties in their context. By the third year about 90 per cent of the area was cropped by local land races. Already influenced by Chipko movement chugging trees to save them from logging contractors) in the region, a save the seed movement emerged, with a number of NGOs and volunteers participating in it. According to the Save the Seed reports they have identified and multiplied some 126 types or varieties (land races) of rice, 8 of wheat, 40 of finger millet, 6 of barnyard millet, 11 of kidney beans, 7 of horse gram, 8 of traditional soyabean and 10 of French beans. They are being grown and used by the farmers. The monoculture encouraged by new technologies is again replaced by Barahanaja (mixed cropping of twelve grain crops).
Sources: V. Singh, 1995 and Kothari, 1997.

---

biodiversity/resource-use systems; (c) possibility of higher financial gains from diversified organic farming compared to the conventional, largely external input-dependent agriculture; and (d) prospects (though quite dim at this stage) of fair pricing of high-value products (e.g. herbs, etc.) from CPRs, and building a value-adding diversification strategy (through processing, etc.) on the components of local biodiversity.

To harness the emerging possibilities through sustained work, the first precondition is to create space for the people's biodiversity management systems in the mainstream work on the subject. Once again, lest the possible provision of space and resources for people-managed biodiversity is treated as a charity, it will be useful to: (a) project its relevance for and usability by the mainstream work under biodiversity conservation strategies, and (b) indicate the approach or methodology for understanding and integrating the

BOX 8.6. CROP DIVERSITY IN THE ECUADORIAN AMAZON: MIMICKING TROPICAL FOREST ECOSYSTEM

Survey data collected in the upper Ecuadorian Amazon show that settler farmers within their new farming systems incorporate the broad elements of polyculture-based farming methods, and practices. The share of gross cropped farm areas (cleared land minus land devoted to pasture and fallow) allocated to each crop varied substantially across sample farmers, suggesting that most farms had a mixed system of landuse.

Accordingly, 16 per cent of the farm area is planted to perennial crops (including mainly coffee, but also cacao, African palm oil, and fruit trees); 5 per cent to annual and semi-annual food crops (including plantains, corn, manioc, rice, vegetable and others); 22 per cent to pasture (which includes areas of fallow or *rastrojo*); and 57 per cent remained in undisturbed forest.

Settler plots in this area of the Amazon mimic tropical ecosystems in at least two ways (a) the great diversity of crops grown gives some protection, as pests are seldom able to build up to destructive proportions on the few isolated plants of each species. Also the closed canopy consisting of some trees left standing and tall crop species such as bananas and papayas reduce losses to pests and weeds; (b) selective burning rotation, intercrossing and shading help reduce losses to pests and weeds. As only relatively small plots are cleared, biological agents can easily enter from the surrounding jungle. Settlers also select for host resistance by using seed and vegetative parts from the most successful crop plants which survive in the harsh environment.

Although it may be premature to draw conclusions abut the long-term sustainability of this agricultural system in north eastern Ecuador, the polycultural system seems to promote greater stability and conservation of biodiversity, in contrast to the rapid turnover of colonists and resource degradation observed in most other agricultural frontiers, where boom and bust economies dominate small farmers' psychology.

Source: F. Pichon, 1996. (Personal communication based on his Ph.D. Work.)

people's biodiversity management systems into the mainstream work on the subject.

CREATING A SPACE FOR PEOPLE'S BIODIVERSITY MANAGEMENT SYSTEMS

The advocacy of people's biodiversity management systems can be supported by projecting their importance in different contexts, as indicated below (see Chart 8.2):

(a) Even when the biodiversity managed by people in 'bush' and 'backyard' is not as rich as the one in the untouched protected areas, the former when aggregated at different levels, may significantly contribute to the global stock of biodiversity. Thus improved local management of biodiversity has clear global gains. This does provide a channel of linking local and global perspectives and action on biodiversity conservation.

(b) The use-driven focus of people-managed biodiversity conservation reflects a dual-purpose strategy of the communities, where conservation for the future is combined with meeting current needs. This represents a type of situation where, often repeated advocacy for win-win approach to environmental management materializes by ensuring not only positive biodiversity conservation outcomes but also yielding tangible social utility.

(c) In the context of increasing recognition of the need for involving people for cost-effective and sustainable natural resource management including biodiversity conservation, the people's existing systems focused to biodiversity in 'bush' and 'backyard' can prove a useful entry point. Understanding of people-managed initiatives and their knowledge systems can serve as a useful input in evolving biodiversity conservation strategy at higher levels.

(d) Furthermore, an understanding and application of dominant features of people's systems (e.g. use-driven biodiversity management) can help evolve conservation approaches readily acceptable to the people to ensure their participation.

(e) The significance of biodiversity conservation in 'bush' and 'backyard' is probably the greatest for mainstreaming biodiversity in agricultural development. The conventional agricultural development approaches, characterized by over-emphasis on limited crops as well as monocropping and disregarding diversified cropping and resource use, can borrow a lot from people-managed biodiversity systems in 'bush' and 'backyard'. Most importantly, the latter are a potential source of varied land races and wild relatives of already used cultivars. People's knowledge and practices can offer some usable insights for developing approaches to diversified and high-productivity agriculture (Altieri et al. 1988).

If these considerations are able to justify greater attention to people-managed biodiversity, the next step should focus on the development and adoption of measures directed to: (a) controlling the factors and processes which are contributing to the rapid decline of people-

244 • *Life on the Edge*

Chart 8.4.   An Indicative List of People's Practices Reflecting Biodiversity
            Management Focus

Land Use
    Extent of CPR (area under natural vegetation)
    Intensity use of CPR—through density of grazing animals plus some
    idea of fodder/fuel pressure
    Area under cropping and cropping intensity

CPR Management
    Their extent
    Arrangements for area protection, usage regulation, etc.
    Oral history of changes including underlying factors

Cropping Patterns
    Major crops and minor crops planted
    Source of planting material
    Extent of intercropping, rotations and other agronomic practices,
    especially using/regenerating local inputs
    Diversity of cropping— number of crops per hectare
    Multiple uses of crops
    Extent of external seed/input, commercialization

Qualitative Dimensions
    Folk agronomy and oral history of change and impacts
    Food systems and agricultural product demands—food preferences, self-
    provisioning
    History and processes of new initiatives promoting BD

---

managed arrangements for biodiversity conservation, and (b) designing and use of methods to understand, document and use the elements
from people-managed systems for strengthening both the national
biodiversity conservation strategy and agricultural development
strategy.

CONTROLLING THE FACTORS AND PROCESSES ADVERSELY
AFFECTING PEOPLE-MANAGED BIODIVERSITY: FOCUS ON
PUBLIC POLICY

Our focus in this context is only on public policies/programmes,
designed and implemented by the national governments. The governments are not only the signatories to the Biodiversity Conservation
Treaty, but they alone have the authority to alter the public policies/
programmes and create an environment conducive to promotion of
people's biodiversity management systems. Chart 8.3 provides a
preliminary structure to list the important public policies/programmes

having unfavourable orientation toward people-managed biodiversity in both habitat and species contexts. They include land policies, agricultural intensification strategies, responses to population pressure and poverty problems, product and resource pricing and trading policies, centralized and top-down approach of public intervention, and biases characterizing the fiscal and infrastructure systems. The recognition and reversal of approaches indicated in Chart 8.2 would constitute the key steps toward initiating new policy and programme strategies directed to people-managed systems. The actual steps to identify relevant policies and programmes and their implementation will be very much location-specific. Nevertheless, a broad approach to them should also form a part of the methodology. Chart 8.5 can help to provide a broad framework for the purpose.

METHODOLOGICAL AND OPERATIONAL ISSUES

TASK-DETERMINED METHODS

Methodology for a task is largely determined by the goals it has to serve. Defined in terms of major and minor goals, the tasks to be served by proposed methodologies would include recognition and promotion of people's biodiversity management systems and their use in national biodiversity conservation programmes with specific focus on agricultural and natural resource development. The broad steps it would involve are:

(a) Accumulation and synthesis of information about the people's systems and its dissemination for awareness generation as well as policy dialogue and decisions. This has to be done in the context of predetermined typologies of situations in terms of agro-climatic and social (especially population density) conditions.
(b) For information generation and analysis, a simple step could be to see through the records or secondary data on variables which could be used as proxies for key aspects of people's biodiversity management.
(c) Statistics on changes in land-use patterns and cropping pattern may give an idea of the changing status of 'backyard' and 'bush' (in aggregate terms) as biodiversity habitats.
(d) The broad understanding provided by the above can be validated with micro-level focused studies using PRA (participatory rural appraisal), etc.

Chart 8.5. Social Dimensions of Biodiversity Management in Agricultural Landscapes

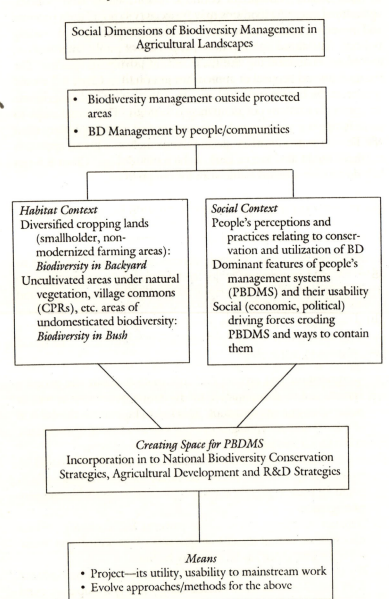

(e) However, to have both qualitative (and to some extent quantitative) assessment of the situation, application of social assessment (SA) methods may prove very helpful. Through this not only can one capture the oral history of changes in a micro-context but can also collect people's views on the factors or processes behind the change, as well as possible ways to control negative consequences of such changes on the people's biodiversity management systems, etc. One can also get a better idea of people's valuation of biodiversity through SA and PRA (IIED 1995).

(f) The use of secondary data as well as synthesis of available knowledge, and use of PRA and SA methods in their respective domains can be applied to any of the areas of inquiry listed below. For example, if government's land policy or agricultural R and D are identified as factors adversely affecting people's biodiversity management, information on them too can be built up through formal statistics (if available) and interactions with relevant communities.

(g) Similarly, information on different aspects (e.g. usability, valuation, changing status, etc.) of the people's biodiversity management systems can be assembled through PRA/SA.

(h) To illustrate the approach, a structure to cover the major aspects is given in Chart 8.4. The data sources and their situation-specific analysis as given in the chart can help in identification of 'good practices' of people-managed biodiversity for incorporation into conservation strategy.

(i) To identify major contexts in the area-specific situations, the broad conceptual framework can be developed using Chart 8.5.

9

Reviving the Social System–Ecosystem Links in the Himalayas*

It is often inferred that present-day society—particularly the policy-makers, planners, and their technical advisers dealing with the mountain regions—is better equipped than the traditional communities in terms of knowledge of ecosystems and their functional dynamics. And yet, it is unable to design and implement a social framework (covering norms and mechanisms to govern a community's approach and interactions with nature) which could more appropriately respond to the imperatives of the natural-resource base. Traditional communities, though innocent of the knowledge of the formal principles of ecosystem (or natural resource system) dynamics, did understand the manifestations of these dynamics, largely in terms of the myriad opportunities and constraints for the community's survival. Consequently, they evolved norms and practices to regulate individual and collective behaviour (*vis-à-vis* nature) as well as technical and institutional mechanisms to support them. This in turn helped in shaping and sustaining positive ecosystem–social system links. Traces of such links, though under severe strain, can still be found in several parts of the Hindu Kush–Himalaya (HK–H) region (or other relatively inaccessible ecosystems), where modern changes disruptive of such linkages are yet to make their full impact.

This chapter deals with the natural resource-friendly traditional patterns of resource use in the HK–H region, their progressive decline and possible approaches to their revival. Map 9.1 presents the topography of the region. The chapter draws on the broad synthesis of inferences and understanding generated by more than fifty studies by different agencies in different parts of the region. The areas include

*First published as 'Reviving Social System–Ecosystem Links in the Himalayas', in F. Berkes and C. Folke (eds), *Linking Social and Ecological Systems*, Cambridge University Press, Cambridge, 1998.

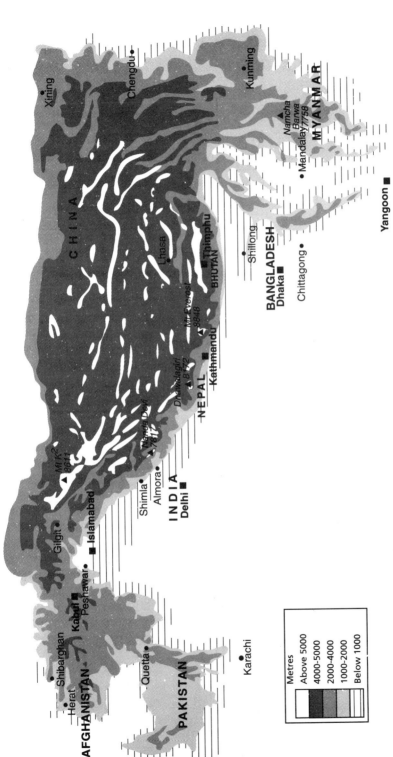

Map 9.1 Topography of the Hindu Kush–Himalaya Region

West Sichuan and Tibet in China; Himachal Pradesh and the hill areas of Uttar Pradesh states in India; middle mountains of Nepal; and the North-West Frontier Province of Pakistan. The primary focus of the studies and operational work was mountain (including hill) agriculture, covering all land-based activities such as cropping, horticulture, animal husbandry, forestry, and their support systems.

In the very early stage of work, those involved in these studies were alerted, both by reviews of existing literature and by field investigations, to several negative trends characterizing agriculture and the overall natural resource situation in the region. These persistent negative changes manifested in the emerging prospects of unsustainability of resource use in most parts of the HK–H region. These trends suggested that present patterns of resource use and production practices (in terms of choice of activities and resource-use intensity) were at odds with the imperatives of the features of natural resources in the region. The relevant key features of mountain areas (termed mountain specificities) included an incredibly high degree of inaccessibility, fragility, marginality, diversity and unique production opportunities with comparative advantage to mountain areas compared to other regions (Jodha et al. 1992). The integrated mountain specificities, providing both opportunities and constraints, represent the 'ecosystem' as people would understand it and respond (adapt) to it. Similarly, the patterns of human adaptations to the circumstances created by mountain specificities, as reflected through resource-use practices and the technological as well as the institutional arrangements supporting them, represent the 'social system' in the context of the present discussion.

Drawing on earlier studies since 1987, inventories were prepared of past and present resource-use systems, production and consumption practices, and technological and institutional measures, including demand management and resource upgrading. The inventories were related to the imperatives of mountain characteristics to assess the degree of match or mismatch between the imperatives of resource features (representing ecosystem) and attributes of the resource-use systems, including their technological and institutional underpinnings (representing social systems). The persistent negative changes, due to mismatches, described as indicators of unsustainability (Jodha 1990b), relate to:

(a) health of natural resources (e.g. increased landslides and other forms of land degradation indicate poor health);

(b) productivity of land-based activities (e.g. persistent decline in crop yields and biomass yields from pastures); and

(c) range and quality of resource management options (e.g. non-feasibility of resource-regenerative agronomic practices, farming–forestry linkages).

Indicators under (a) and (b) represent the disruption of ecosystems, while those under (c) represent the disruption of social systems; and, finally, all three categories of indicators represent the process of one disruption reinforcing another.

IMPERATIVES OF MOUNTAIN CONDITIONS AND HUMAN RESPONSES

Under the traditional systems, people understood and responded to ecosystems in terms of constraints and opportunities generated by the specific features of their natural resource base. The feature of biophysical resources or the circumstances created by them thus shaped the social perspective, i.e. values, norms and mechanisms for individual or collective interactions with nature (Jochim 1981). In mountain areas, human responses were shaped by mountain specificities. Chart 9.1 describes the situation in relatively broad terms. The prevalence of traditional practices declines as one moves from relatively remote to more accessible villages. The consequences of these changes were quite visible, and in most cases people recognized them as part of their concerns and their vision of the future for their children.

Although our context is mountain areas, the formulation and analysis presented below may have general applicability to many traditional or indigenous communities in semi-isolated situations. Accordingly, Chart 9.2, which summarize's the dynamics of ecosystem–social systems linkages and tries to address the issues, indicates (a) the nature-dominated key objective circumstances under which the small and relatively isolated communities lived and managed their natural resources—through improving accessibility, local resource dependence, etc.; (b) the key driving forces and factors which shaped societal responses to the said objective circumstances; (c) broad social responses in terms of concerns and adaptation strategy; (d) technological and institutional mechanisms evolved and adopted for implementing this strategy; and (e) consequences of (b) to (d) in terms of evolution of nature–society interactions and sustainability of resource use. These features of traditional systems are contrasted

Chart 9.1. Mountain Resource Characteristics, Their Imperatives, Objective Circumstances and Driving Forces Behind Human Response

Resource Features and Objective Circumstances	Imperatives—Driving Forces	Response, Resource-Use Practices
Inaccessibility (Caused by physical, terrain factors) imposing high degree of isolation, poor mobility, and limited external linkages, semi-closed	Survival strategies with direct and total dependence on local resources and high stake in their protection, regulated use and regeneration; local control of local resources, culture of self-management, evolution of systems from below based on closer proximity and knowledge of resource base	Ecology-driven resource management using conservation and protection technologies, and institutional arrangements, evolved with closer feel of the resources and enforced through local autonomy and control of local resources; rationing of demand pressure on resources, and restricting extraction levels in keeping with subsistence needs
Fragility (Caused by biophysical, topographic, edaphic characteristics) Vulnerable to irreversible degradation with small disturbance, restricting usage options, intensity levels	High risk of rapid resource depletion due to intensification, hence, measures to balance extrication and conservation of production base; narrow range of production options (only land extensive uses)	Technologies and usage practices combining intensive and extensive uses of natural resources; provision of institutional arrangements (e.g. common-property resources) against overextraction of fragile/marginal resources, spatially and temporally differentiated resource use systems/rationing; knowledge and capacity-based resource upgrading (e.g. by terracing, agroforestry, etc.)

Diversity (Created by huge variations in biophysical features and elevations at shorter distances) Creating opportunities for diversified interlinked production/consumption activities	Local knowledge, skill and capacity-based diversification of resource use as a key element of survival strategies; sustainable productivity, health of natural resource base	Spatially and temporarily diversified and interlinked activities with varying levels of intensification; diversification of demands to match the diversity of products and supplies, especially in a semi-closed situation
Niche (Created by unique agro-climate, biophysical situations) Imparts comparative advantage to mountain areas in some activities and products (forests, horticulture, herbs hydropower, etc.)	Potential for trade-based external linkages restricted by levels of knowledge, capacities to harness, etc.	A limited range of diversified activities directed to petty trading to supplement subsistence activities: local niche, demand and extraction facilities/capacities as key factors governing the exploitation of niche
Implication	Adherence to two-way adaptation process	Ecology-driven systems of resource use conducive to sustainability (under low pressure of population and external demand)

Source: Based on synthesis of accounts of concrete situations described in over 45 studies in mountain areas covering Nepal (18), China (15), India (7), Pakistan (3), Bhutan, Bangladesh and Myanmar (1 each) as synthesized by Jodha and Shrestha (1994).

Chart 9.2. Factors and Processes Associated with the Nature–Society Interactions under Traditional and Present-day Systems of Resource Use in Mountain Areas

Traditional Systems	Present-day Systems
A. *Basic Objective circumstances*	
Greater degree of inaccessibility, isolation and semi-closedness of systems; poor mobility and external linkages, etc. creating total and exclusive dependence on local resource base and high concern for its health and sustainable use	Greater physical, administrative and market integration of traditionally isolated areas/communities with the dominant, mainstream systems, reducing critical dependence of the former on local resources and hence the degree of their stake in the conservation of local resources
B. *Key driving forces/factors generated by (A)*	
Social survival/welfare strategies totally focused on local, diverse, fragile resources	External linkages-based diversification of sources of sustenance, welfare and development reducing the extent of critical stake in local resource maintenance
High collective stake in protection and regeneration of local natural resources	
Functional knowledge and closer understanding of limitations and potential of resources due to closer proximity and access to resources, little gap between resource user and resource itself	Role of functional resource knowledge marginalized due to imposition of generalized approaches from above for local resource management; wider gap between resource users and decision makers
Autonomy, local control over local resources (due to absence of external impositions)	Erosion of local resource control, autonomy following the extension of mainstream, legal, administrative, fiscal arrangements to formerly isolated areas
Low population pressure as permitted by biophysical constraints	Rapid demographic changes
C. *Social Responses (concerns and adaptations) dictated or facilitated by (B)*	
Adoption and enforcement of production/ extraction systems adapted to natural resource features through	Greater role of demand-driven measures leading to resource-use intensification, overexploitation with greater extractive

diversified usage, controlled usage-intensive; regenerating, upgrading, developing the resources, depending on capacities and needs

Controlling or rationing the demand pressure on resources through social and institutional sanctions, collective sharing, recycling, out-migration, etc.

D. Mechanisms and means to execute social responses

Collective evolved site- and season-specific norms of resource use facilitated by direct access and proximity to resources and little gap between decision makers and resource users

Site, season, product and resource component-specific folk-technologies evolved over the generations facilitated by functional knowledge and close proximity to resource base

Formal/informal institutional arrangements guiding broad approach to resource management, access and usage regulation, facilitated by group action or community participation, and autonomy and local control over local resources

E. Consequence: Ecology-driven natural resource management systems

Evolved by the communities having high stake in sustainability of the resource base

Facilitated by functional knowledge of resources, close proximity to resources and community control over the local resources

capacities and technologies

Increased role of (unregulated) external demands, which are insensitive to local resource limitation

Resource upgrading measures more generalized and less location specific

Largely externally evolved generalized rules guiding resource use, framed by legal and technical experts with little concern for local resource users' perspectives and limited knowledge of site-specific situations

High science-based modern R&D as a source of technologies, ignoring rationale of traditional practices; ignore local resource perspectives

Institutional interventions evolved and designed for incomparable situations extended to these areas as a part of agricultural, rural development, etc.

Resource usage system driven by uncontrolled pressure of demand:

Developed by experts without local participation

Enforced (rather unenforced) by formal state machinery

Source: Based on a synthesis of accounts of concrete situations described in over 45 studies in mountain areas covering Nepal (18), China (15), India (7), Pakistan (3), Bhutan, Bangladesh and Myanmar (1 each) as synthesized by Jodha and Shrestha (1994).

with the changes following rapid population growth as well as the physical, administrative and market integration of hitherto semi-closed/isolated systems or areas within the mainstream society. Integration, despite its various gains to mountain areas, has adversely affected the traditional resource management system (Banskota 1989; Collier 1990).

Chart 9.2 is relatively self-explanatory, but its key points may be briefly summarized. The community's biophysical environment, characterized by a high degree of inaccessibility, imposed a certain degree of isolation and necessitated self-sufficiency. In the absence of effective outside linkages, their sustenance and welfare depended totally or crucially on the local resources. This forced them to adapt their requirements, etc. by demand rationing as well as their resource-use systems to the limitations and potential of local resources, rather than attempting to manipulate or over-exploit resources to satisfy uncontrolled human needs. They had a high stake in the health and productivity of local natural resources. Close proximity to natural resources, local control of resources, intimate functional knowledge about them (again largely because of the closedness of the system), and lower pressure of population helped the communities to evolve folk-technologies and institutional arrangements, and to enforce them without external interference, for the protection, regeneration and regulated use of their resources. In the process, attitudes and norms of socio-economic behaviour, which had gradually evolved for the use of biophysical resources of the community, helped in linking social system with ecological systems to ensure sustainable use of resources in a subsistence context (Guillet 1983).

With the changed circumstances associated with the increased integration of hitherto isolated areas with the mainstream areas (as well as their population growth), ecology-driven social responses and resource management systems faced a rapid decline. While the integration of isolated/indigenous areas may be justified on several grounds, the process involved (using the norms and procedures characterizing the mainstream, i.e. prime land, industry and market-dominated areas) has marginalized the areas and communities in question. As a result, while the biophysical context remained largely unchanged, the socio-economic circumstances in these areas have changed rapidly (Bjonness 1983; Ives and Messerli 1989). Chart 9.3 summarizes various changes, and their impact on the resource base, production flows and resources practices. Next we shall focus on one major change, that relates to population.

POPULATION GROWTH

The negative impact of integration with external systems was accentuated by rapid population growth in these hitherto semi-closed mountain areas. While population growth is largely an internal change, the external linkages have also played an important role in the process. First, growth of life-saving health facilities available from the outside, although much less than required, has brought down the death rate even in remote mountain areas. Second, the possibilities of external relief support during crisis and scarcity, although less than adequate, have eroded internal demand-rationing measures (including traditional methods of population control such as the practice of the elder son in the family becoming a Buddhist monk, and not marrying, a practice still prevalent in parts of the Himalayas). Third, and most important, the demographic impact of external linkages took place in terms of qualitative changes in the population. Growth of individualistic tendencies, and disregard or indifference towards social sanctions and collective action, which were key to traditional resource management, have emerged following the establishment of external links. This represents a rapid decline of 'social capital' in mountain areas.

Thus the local-level ecosystem–social system linkages were disrupted by the emergence of a complex of internal (population) and external driving forces (e.g. interventions through market and state). The pressure on resources, encouraging their over-extraction, increased. Unlike in the past, the total dependence on local resources ceased to be a key driving force to sustain people's stake in resource stability. The positive effects of local autonomy, control over local resources, close proximity to resources, functional knowledge of resources and social cohesiveness, all of which in the past helped in the development of technologies and institutional responses, weakened. Integration also meant the imposition of irrelevant techno-logical and institutional measures from the outside (Banskota and Jodha 1992a).

Because of these rapid and major changes, the local communities were left without sufficient lead-time or control over their resources and community affairs to amend their age-old coping strategies or to evolve new ones. Furthermore, they did not have the capacity or even incentives to resist the internal and external forces released by their integration with the stronger, external systems and the unprecedented growth of population. Their knowledge systems, social sanctions, collective sharing system, etc. became less effective or less feasible

Chart 9.3. Negative Changes as Indicators of Emerging Unsustainability in Hindu Kush–Himalaya Region

Visibility of change	Change Related to[a]		
	Resource Base	Production Flows	Resource Use Options
Directly visible changes	Increased landslide and other forms of land degradation; abandoned terraces; per capita reduced availability and fragmentation of land; changed botanical composition of forest/pasture; reduced water flows for irrigation, domestic uses, and grinding mills	Prolonged negative trend in yields of crop, livestock, etc.; increased input need per unit of production; increased time and distance involved in food, fodder, fuel gathering; reduced capacity and period of grinding/saw mills operated on water flow; lower per capita availability of agricultural products, etc.	Reduced extent of following, crop rotation, intercropping, diversified resources management practices; extension of plough to submarginal lands; replacement of social sanctions for resource use by legal measures; unbalanced and high intensity of input use, subsidization
Concealed by responses to changes	Substitution of: cattle by sheep/goat; deep-rooted crops by shallow-rooted ones; shift to non-local inputs Substitution of water flow by fossil fuel for grinding mills; manure by chemical fertilizers	Increased seasonal migration; introduction of externally supported public distribution systems (food, inputs): intensive cash cropping on limited areas	Shifts in cropping pattern and composition of livestock; reduced diversity; increased specialization in monocropping; promotion of policies/programmes with successful record outside, without evaluation
Development initiatives,	New systems without linkages to other diversified activities and regenerative processes; generating excessive dependence on outside	Agricultural measures directed to short-term quick results; primarily productions (as	Indifference of programme and policies to mountain specificities; focus on short-term gains; high

etc., i.e. processes with potentially negative consequences[b]	resource (fertilizer/pesticide-based technologies, subsidies); ignoring traditional adaptation experiences (e.g. new irrigation structure); programmes focused mainly on resource extraction	against resource) centred approaches to development; service-centered activities (e.g. tourism) with negative side-effects	centralization; excessive, crucial dependence on external advice ignoring traditional wisdom; generating permanent dependence on subsidies

[a]Most of the changes are interrelated and they could fit into more than one block.

[b]Changes under this category differ from the ones under the above two categories, in the sense that they are yet to take place, and their potential emergence could be understood by examining the involved resources-use practices in relation to specific mountain characteristics. Thus they represent the 'process' dimension rather than consequence dimension of unsustainability. It is based on data or description by over 45 studies from Nepal (18), China (15), India (7), Pakistan (3), Bhutan, Bangladesh and Myanmar (1 each) as synthesized by Jodha and Shrestha (1994).

Sources: Adapted from Jodha (1990b), Jodha and Shrestha (1994).

and less attractive (especially to the younger generations) in comparison with externally supported arrangements. In the final analysis, the
whole complex (type and nature) of driving forces and patterns of
responses to them changed. The net consequence was the emergence
of what could be considered as indicators of unsustainability (Chart
9.3). Viewed in the context of the thematic framework of nature–
society interactions mentioned earlier, these negative changes reflect
a complex of disruptions, where social systems tend to behave
independently of the imperatives of ecosystems. The two-way
adaptation process is converted into a one-way adjustment, whereby
resource manipulation and extraction are over-stretched to meet
increasing human demands, rather than adjusting the latter to the
limits of resource availability. This led to the breakdown of resource-
regenerative, diversified production systems; indiscriminate resource-
use intensification (often maintained through a high level of chemical,
biophysical and economic subsidies); and the depletion of resources
(Jodha 1995a).

RESTORING ECOSYSTEM–SOCIAL SYSTEM LINKS: EXPLORING THE POSSIBILITIES

By looking closely at the traditional systems, one can identify some
key elements of the circumstances and processes which were
responsible for positive ecosystem–social system links. These include
the processes which generated community concerns, commitments,
incentives and facilities for the protection and regulated use of natural
resources; enhanced community capacities (both technical and
institutional) to respond appropriately to biophysical circumstances
through combining production and protection-centred measures; and
motivated and facilitated enforcement of measures that helped in
adapting community needs to resources rather than manipulating
and over-extracting the latter to meet unrestrained demands. Accordingly, one can identify three elements which individually or jointly
strengthened the ecosystem–social system links and contributed to
the natural resource-friendly traditional management systems: (a) a
total dependence-driven stake in the protection of natural resources;
(b) close proximity and a functional knowledge-driven approach to
resource use; and (c) local control-determined sanctions and facilities
governing resource use. Also, the smaller populations and greater
social cohesiveness of traditional societies were the major facilitators

of the above responses. The following discussion elaborates on these three points with a view to exploring some possibilities of reinstating these elements, or identifying their present-day functional equivalents, as parts of an incentive structure to facilitate positive social approaches essential for sustainable resource management in mountain areas. The main points are summarized in Chart 9.4 and expanded upon in the text.

The key premise behind this exercise is that even though semi-isolation and other key objective circumstances characterizing traditional communities formed the basis of community stakes and sensitivity towards natural resources, the feasibility and viability of ecosystem-friendly attitudes and practices are not confined to small and isolated groups. By changing the forms of their manifestation and their operating mechanisms, these elements can be integrated into any resource management system and can prove effective in any context.

In the relatively less accessible mountain areas, exclusive or total dependence for sustenance on local resources was the key incentive behind communities' concern and the follow-up actions which led to protection and regulated use of their natural resources. To reiterate, activities from the combining of production and conservation measures to the rationing of demand, as well as adherence to social sanctions regulating resource use, can be easily linked to the uniqueness of traditional societies and their incentive systems. Reinstating such incentive or disincentive systems is neither desirable nor feasible now. Nevertheless, it is possible to explore other approaches to strengthen the dependence-driven community stake in the natural resource base. One such possibility involves change of the product context of dependence. At the local community level, the traditional security of sustenance can be substituted by the security of the biophysical niche, reflected through the comparative advantage (due to specific high-value options) potentially available to the community through physical and market integration of their area with the mainstream economy. There are cases of transformed areas in the HK–H region and other mountain regions where such niche-based gains have worked as new incentives for the protection and regulated use of the natural resources base by the communities.

For example, in Ningnan county (West Sichuan, China), where sericulture has recently become a lead activity with a high pay-off and comparative advantage to the area, communities attempt to manage and protect hill slopes, shrubs and waterflows on a priority basis

Chart 9.4. Possibilities of Reorienting Current Resource-usage Systems in Mountain Areas by Incorporating Elements from Traditional Systems

Circumstances, Driving Forces, Response Mechanisms Characterizing Traditional Resource Use System	The Element of Traditional Systems with Scope for Revival, Reorientation and Substitution in the Present-day Context
Total dependence-driven community stake in natural resource base Key factor: Almost total and exclusive dependence on local resource for survival (in a semi-closed, isolated subsistence-oriented context) inducing protection, regeneration and sustainable use of resources; the process was complemented by close proximity and functional knowledge of the resources which sharpened community's perception and diagnosis of resource situation Infeasibility (and undesirability) of this in the changed context of reduced isolation and access to external sources, etc.	*Rediscover areas of total/crucial dependence as sources of community stakes in natural resources* Change of product/service—context of stake, i.e. substituting (traditional) sustenance security by security of 'niche' (high pay-off products/services with comparative advantage to the local communities) and use them as lead sector influencing overall natural resource management (e.g. horticulture- or tourism-led initiatives in mountains)
Close proximity, direct access to resources, and their functional knowledge Generated understanding and sensitivity to resource situation, and its variability; helped in developing relevant folk-technologies; encouraged institutional arrangements for resource-use regulation (intensity, diversification, common-property regimes, etc.); reduced gap between resource user and decision maker, producer and produce consumer, and helped in regulating pressure on resources	*Functional substitutes for close proximity and firsthand knowledge of resource base* Focus on sensitivity in place of proximity to resource situation, as the latter contributed to the evolution of resource management measures mainly by generating sensitivity towards the resources; re-orient, sensitize policy-makers, development agencies (even market forces) to make them understand imperatives of mountain resources and act accordingly

After integration with the mainstream (external system), leading to distance between the resource and resource planners; multiplicity and diversity of resource users and pluralization of perception of stakes and marginalization of traditional knowledge systems, the physical proximity-dependent approaches are not feasible.

Autonomy and community command over local resources

A product of isolation or semi-closedness of areas/communities that facilitated effective community ownership of resources; helped in designing and enforcing resource-use regulations; reduced gaps between decision makers and resource users; encouraged community participation and group actions; and helped in resource-use rationing, collective sharing, resource recycling, and protection against pressure of (possible) external demand

The above features and functions have vanished or weakened with the integration of isolated areas with the mainstream economy and demographic changes. Revival of the system conflicts with 'centralization' and top-down approaches of the mainstream decision makers, governed more by the interests and perspectives of the mainstream

Evolve feedback mechanisms (about resource situation) by involving local communities as a substitute for instant feedback provided by physical proximity in the past

Restoring community management of local resources

Build on the emerging trends toward decentralization, community-based development, participatory and bottom-up approaches to development. Experiences of user group and NGO initiatives, etc. can help provide functional substitute arrangements for traditional group action through community control of resources

Take leads from successful experiences, community forestry, community irrigation systems and other grassroots level participatory initiatives, facilitate their replication and mainstreaming; sensitize decision makers to new possibilities

because it helps in strengthening sericulture activities. Other examples include Himachal Pradesh (especially the apple zone) in India, or the Ilam district of Nepal, where multiple new activities are sustained through better management of natural resources in general. Where mountain environment and landscape have become major tourist attractions (including the Alps and pockets of the Himalayas), the same logic of incentive-through-stake has helped improve the management of natural resources by the communities. There are many such examples where a stake in the lead sector/lead activity (due to biophysical and economic interlinkages) has induced and initiated a process of better management of overall resources by the communities (Jodha and Shrestha 1994). Exceptions occur where niche is identified and harnessed (or extracted) without involving the local communities.

Yet further examples of the revival of the community stake in natural resources through common perception of needs, are the forest-user groups in Nepal and Indian. In most such cases, having reached the threshold local level of scarcities of fodder, fuel, etc. due to degradation of forests resulting from external and internal demand pressures, communities (with and without the help of NGOs) have revived systems of collective protection and regulated use (Poffenberger and McGean 1996). The same applies to the revival of community irrigation systems in the hills of Nepal, Pakistan and India.

The key factor in all these success stories is the user groups' common perception of stakes and collective action. This represents a step towards the rehabilitation of communal integrity or the rebuilding of social capital. As a total picture involving natural resources and communities, this change manifests the revival of ecosystem–social systems linkages.

PHYSICAL PROXIMITY AND FUNCTIONAL KNOWLEDGE OF THE NATURAL RESOURCE BASE

In traditional systems, while the community's stake in its natural resources was an important driving force in shaping society's approach to ecosystem and inducing a conservation-oriented resource management system, an equally important role was played by site- and season-specific functional knowledge of the resources. This in turn was gained through proximity and access to resources. Physical proximity and functional knowledge thus had a mutually reinforcing role in enhancing and sharpening people's perceptions of their stake in the better management of their natural resources. This resulted in the development of resource-use systems, the generation of folk-technologies, the creation of demand-rationing measures, and the institutional arrangements to facilitate their adoption and enforcement. The balancing of extensive and intensive types of land uses, various forms of resource-use diversification and flexibility, resource regenerating, recycling practices, methods of resource upgrading (i.e. by terracing), seasonal and periodic restrictions on product gathering from the village commons, are some of the concrete instances where a better understanding of resource features and the availability of a longer lead-time for informal experimentation helped communities.

Absence of any gap between decision-making and the actual use of resources, as well as between the resource user (i.e. producer) and the product users (again facilitated by proximity and access), helped in encouraging flexible approaches to resource management to meet site- and season-specific differences and contingencies. They also helped in adjusting people's requirements (i.e. animal grazing intensity, seasonal collection and use of food, fuel and fodder) to the availability and potential of the resource base. Restriction on the collection of specific products during specific periods or from specific areas, and enforcement of grazing rotations by local communities in some villages, even today, are illustrations of such adjustments.

Unlike in the past, the greatly modified present-day resource-use situation is characterized by: (a) a wider scatter of users of the resource products (due to market integration) and an equally wide gap between the producer (resource user) and the product consumer (e.g. the final user of herb and horticulture products and hydropower from mountain areas); (b) the dissociation between usership and ownership of resources (due to growth of absentee landlordism in many areas); (c) the dissociation of decision-making agencies (legal, fiscal and administrative authorities) and resource-using groups (i.e. the farmer or the community); and (d) the distance and differences between technology developers and technology users. These circumstances restrict the scope for reinstating and strengthening resource management practices which are closely tied to physical proximity, direct involvement and accessibility to the resource base, and first-hand knowledge. To take fuller advantage of knowledge and understanding of the resource base, to reorient an ecosystem-friendly social approach, and to design and implement relevant usage/management systems, however, it is not necessary to recreate the traditional situation characterized by semi-closedness and close physical proximity to resources. In the present-day context, these goals can be achieved through better means of information acquisition, verification and synthesis, as well as communication and dissemination.

To benefit from first-hand experience of the field situation and accumulated traditional knowledge about resources and resource-use systems, there are well-tested methods of involving local communities, e.g. through rapid rural appraisal/participatory rural appraisal (RRA/PRA), in the processes of collection, analysis and utilization of information. Such information can help create sensitivity towards natural resources among diverse stake-holders, from policy-makers to urban consumers.

The key constraint is that these means have been used only by experts from outside. They have not been utilized for building sensitivity and understanding of mountain resources or conditions. The focus of policy and programme interventions—be they agricultural research and development or integrated rural development— has lacked mountain perspective, i.e. understanding and incorporation of imperatives of mountain specificities in the conception, design and implementation of development and welfare activities (Jodha 1995a). If these interventions are seen as part of broader social systems characterizing and influencing mountain areas, they once again reflect the rapidly vanishing links between social systems and ecosystems and their co-evolution in mountain areas. A first step towards filling this gap can be taken by initiating a process directed towards the following.

(a) Sensitization and reorientation of the decision-makers to create a policy environment friendlier to mountain conditions.

(b) Involvement of the local communities in decision-making and actions relating to local resources, to ensure the relevance of interventions to the field situation.

(c) Recognition and utilization of traditional knowledge systems by the formal research and development of agencies engaged in the development technologies and policies for these areas.

(d) Reorientation of the whole process of project planning, designing and implementation by making it a bottom-up approach involving local communities and user groups.

The implementation of these and related measures, in our thematic context, implies overhauling the social systems to enhance their ability to link with the ecosystems. In a less integrated form, these steps already form the mandate of several NGOs and donor-supported activities in many countries.

CONCLUSION: COMMUNITY CONTROL AND RATIONING RESOURCE USE

It is becoming increasingly clear that the management and protection of local-level resources through state agencies such as forest departments, are becoming progressively more difficult and costly. On the other hand, the involvement of local communities in local resource management has improved the situation in many areas (Krishna et al. 1997; Poffenberger and McGean 1996). Awareness and mobilization of local communities, for their rights and resources, enhanced through NGO activism, are also emerging features of communities, even in less accessible areas. The successful negotiations of forest-user groups and community irrigation groups (helped by NGOs) to acquire control of resources in countries like Nepal, India and Pakistan in the HK–H region, are one case in point.

These positive developments assume, however, the building of social cohesion and the mobilization of communities based on shared perception and collective action. Such a social transformation may face several hurdles in the contemporary context. In addition to population increase, qualitative changes in mountain populations, reflected through rapid erosion of community cohesion, weakening of the culture conducive to group action and collective sharing, rapid growth of individualistic tendencies and economic differentiation of communities, may obstruct the effective use of restored community authority over local resources for regulating resource use. Thus, the rebuilding of the social system is a prerequisite for developing or restoring effective linkages with the ecosystem.

Finally, the internal weakness of present-day village communities (constraining community initiatives for resource-use regulation) may be complemented by external forces generated by market and political economy, as manifested through a range of fiscal and pricing arrangements. Over-extraction of mountain resources (disregarding the imperatives of ecosystems), driven by the above forces, may continue despite increases in regulatory powers at the community level. This calls for a gradual process of change focused on the following steps.

(a) Making constant efforts for the greater involvement of communities and user groups supported by NGOs for the planning and implementation of resource management initiatives, use of resource regenerative technologies and regulation and resource use (Daly and Cobb 1989).

(b) Taking the lead from the successful cases of participatory decentralized resource management projects and focusing on their replication and mainstreaming (Poffenberger and McGean 1996).

(c) Helping build capacities and incentives for local communities to adapt to the changed circumstances and revive traditional practices for resource management in the changed contexts.

(d) Introducing different norms for mountain product/resource pricing, reflecting their true worth or environmental cost by building on the conceptual leads provided by recent thinking in this area (Munasinghe 1993).

(e) Introducing biophysical measures for compensation for resource extraction, e.g. reforestation or planting the same type of tree when a tree is cut for the timber market (Jodha and Shrestha 1994).

The foregoing suggestions for creating present-day functional equivalents of traditional circumstances are indicative of the new possibilities. Their design and implementation presume, however, the fulfilment of several preconditions, including the commitment of decision-makers and community-specific preparations.

10

Poverty and Environmental Resource Degradation
Alternative Explanation and
Possible Solutions*

The seeming link between poverty and environmental resource degradation (P-ERD), taken for granted in many developing countries, is emphasized so frequently and vehemently that it looks like a concerted effort to hammer a stereotype. This not only diverts attention away from the several basic issues involved in the process (Panayotou 1990; Metz 1991), but prevents recognition and analysis of simple field-level observations. For instance, there is widespread evidence that in many areas (villages) currently facing severe environmental resource degradation, the resource users in the past were poorer compared to today and yet natural resource degradation was consciously prevented. Similarly, in the current context in many areas the contribution of richer groups towards resource degradation is greater compared to that of the poor. The explanation for such paradoxes lies in the nature and extent of the community's stake in the health and productivity of its environmental resources and institutional and technological mechanisms at its command to safeguard the stakes. Dilution or disintegration of the community stakes and erosion of grassroots level mechanisms to protect and enhance these stakes constitute the fundamental reason behind environmental resource degradation, irrespective of the poverty or riches of the communities. This critical factor guiding the people's approach and resource-use decisions is largely ignored by the generalized mainstream view that links environmental resource degradation

*First presented as a paper, 'Poverty–Environmental Resource Degradation Links: Alternative Explanations and Possible Solutions', at the India-50 Conference at the University of Sussex, Brighton, 1997. This paper was subsequently published in *Economic and Political Weekly*, vol. 33 (36–37), 1998.

to poverty. Consequently, one tends to focus on proximate causes (e.g. poverty) rather than the key driving forces dictating the behaviour of the poor.

QUESTIONING THE PREMISES BEHIND POVERTY–ENVIRONMENTAL RESOURCE DEGRADATION (P-ERD) LINKS

The lead line of reasoning behind the P-ERD view is that poverty (and scarcity) cause desperation, which in turn promotes over-extraction of resources leading to resource degradation and still greater extent of scarcity and poverty, which further accentuates this cycle. This fairly convincingly explains the dynamics of resource degradation and the role of poverty therein. But a major limitation of this formulation relates to the specific assumptions about the approach of the poor to natural resources and their resource-use behaviour. The implicit key premises underlying the formulation that concentrates on poverty as prime mover of environmental resource degradation may be stated as follows:

(a) Over-extraction of resources is the only and preferred means of sustenance the poor know.
(b) The poor are ignorant of both the limitations of their environmental resources and the consequence of their extractive usage practices.
(c) The poor have little stake in the health and productivity of their natural resources.

All these premises can be easily inferred from the current pattern of natural resource use in many poor areas. Our contention, however, is that these are only manifestations of erosion of the past arrangements at grass-roots level, where the situation and behaviour (i.e. decisions and actions of the poor) were quite opposite to the ones implied by the above premises. This we shall illustrate with the help of situations studied in the fragile resource zones, including dry tropical plains and mountain areas in South Asia.

INFORMATION BASE AND BOUNDARIES

The empirical base of the discussion is provided by fairly sustained work for nearly a decade in each, the arid and semi-arid tropical parts of India and the Himalayan mountain region covering different countries. Though mountains and dry plains are very different ecosystems with vast differences in various respects, our argument and analysis are built on their similarities (e.g. their common constraints), when compared with high-potential, prime-land agricultural regions such as the Gangetic Plains (or even valley bottoms in mountains and well-watered pockets in dry tropics). Accordingly, both mountains and dry plains (for different reasons and to different degrees) suffer from fragility and marginality of biophysical resources, historically less accessible and isolated from the mainstream economies/societies; and the rural communities' sustaining themselves through adaptation to harsh biophysical environments without dependable and effective external links on an extensive scale (Jodha 1995a). These broadly similar objective circumstances shaped the positive ecosystem–social system links, which helped in preventing natural resources degradation while using them (Prakash 1997; see Chapter 9). Our argument is based on the synthesis of inferences from different studies relating to broad and typical situations rather than statistical accounts of specific administrative units.

Finally, the term natural or environmental resources used here, covers both crop lands and non-crop lands (forest, pasture and other common lands). This is both because of the organic linkages or complementarities between different categories of land uses, and the central role of 'resource use diversification' in the communities' adaptation strategies in the fragile resource zones. Nevertheless, while discussing environmental or natural resources degradation, our direct focus is mainly on community resources or common pool resources. The crop lands enter the picture indirectly due to organic links between different land-based activities (e.g. crop land–common land complementarities, farming–forestry linkages, etc). Furthermore, the farmers' collective resource management practices extend to crop lands as well. The term natural resource degradation covers deforestation, watershed destruction, loss of biodiversity, loss of soil fertility, soil erosion, ground water depletion, over grazing, fuel and fodder shortages, etc.

WHY CHOOSE FRAGILE RESOURCE ZONES?

In the current context, the chosen areas not only belong to the category of poor areas but are faced with the rapid human-induced degradation of environmental or natural resources (two terms used interchangeably in this essay). Furthermore, the past situation of FRZ, in terms of ecosystem–social system links (i.e. resource users' approach and behaviour) contrasts sharply with the present situation. Finally, as elaborated below, these areas have extremely high potential for both rise of poverty and decline of natural resources.

In the fragile resource zones, due to their biophysical features such as fragility, marginality, low accessibility, internal resource heterogeneity, etc., there is a strong inherent potential for poverty promoting processes (Jodha 1995a). The factors and processes historically associated with prosperity and richness, such as resource-use intensification, risk-free range of high-productivity options, gains of scale, generation and reinvestment of surplus, gains from external exchange, etc. are extremely limited. The people have to live with limited, high-risk, low-productivity options. The mainstream economic and political systems generally found them unattractive and ignored them as marginalities. Thus nature (and mainstream economy as well) generated high poverty prospects for these areas. Destined to be poor, as per the P-ERD formulation, these areas should historically have extreme degree of resource degradation. Furthermore, because of fragility and marginality, these areas are more prone to (often irreversible) natural resource degradation following even a small degree of resource-use intensification, which again as per the P-ERD reasoning, should be unavoidable because the poor through inappropriate intensification over-extract the resources.

Thus poverty of the people and fragility of natural resources make them potentially an ideal place for operation of P-ERD links. The failure of this potential to materialize in the past, however, encourages one to question the over-emphasis on P-ERD links. Also, the understanding of the reasons behind non-working of PERD links can provide useful insights to evolve options for breaking the vicious cycle of poverty–resource degradation–poverty implied by the P-ERD theme. To facilitate this understanding, we can have a quick look at the traditional systems of resource use, based on collective stakes and mechanisms to protect and enhance these stakes in the areas under review.

THE PAST AND THE PRESENT IN FRAGILE AREAS

Here we describe some features of traditional resource-use systems which have direct relevance to the resource-use behaviour of the poor. The objective is (a) to indicate the grassroots-level institutional arrangements, which helped in balancing the protection and extraction of resources to meet sustenance needs, and (b) to reflect on the processes and factors leading to erosion and marginalization of these arrangements.

Chart 10.1 summarizes the inferences from different studies in mountains and arid, semi-arid tropical plains on aspects central to natural resource use in these areas. Accordingly, most of the communities in the relatively isolated, fragile and marginal resource areas, faced with limited, high-risk, low-productivity options; and limited and undependable external linkages, had to evolve their sustenance strategy through adaptation to the limitations and potentiality of their local natural resource base. They included seasonally and spatially diversified and interlinked land-based activities (farming system, common property resources, etc). The key features of the adaptations were:

(a) Near-total dependence on local natural resource base (NRB) leading to community's explicit realization of strong links between their sustenance and protection and productivity of their NRB. Despite internal inequities and occupation-specific differences in gains from NRB, everyone's close dependence on local resources created an integrated collective stake in their natural resource base (Berkes 1989; Leach et al. 1997; see also Chapter 9).

(b) In the context of relative isolation and small size of the rural communities, the latter's physical proximity to their environmental resources imparted better knowledge and understanding of limitations and usability of their NRB (Bjonness 1983). This not only helped in developing folk technological practices to protect and regenerate the resources while using them, but also facilitated the creation of locally enforceable range of regulatory measures to guide use-intensity of resources, and periodic contributions (labour, etc.) towards investment for upkeep and development of the resources (Arnold and Dewees 1995; see also Chapter 5).

(c) Most importantly, enforcement of the above measures was facilitated by social sanctions, group action and, in some cases, feudal arrangements. The ultimate source of strength for enforcement of these arrangements was local autonomy or local control over local resources and local affairs; and the resource

Chart 10.1. Factors and Processes Associated with Community Approaches and Usage of Natural Resources in Fragile
Resource Zones under the Traditional and the Present Systems

Traditional Systems and Approaches	Present-day Systems and Approaches
A. *Structural and Operational Contexts*	
Poor accessibility, isolation, semi-closedness; low extent and undependence external linkages and support; subsistence oriented small populations;	Enhanced physical, administrative and market integration of traditionally isolated, marginal areas/communities with the dominant mainstream systems at the latter's terms; increased population
Almost total or critical dependence on local fragile diverse natural resource base (NRB)	Reduced critical dependence on local NRB; diversification of source of sustenance
Bottom line: High collective concern for health and productivity of NRB as a source of sustenance	*Bottom line:* Reduced collective concern for local NRB; rise of individual (extractive) strategies
B. *Key driving forces/factors generated by (A)*	
Sustenance-driven collective focused on local resource	External linkage-based divesification of sources of sustenance (welfare, relief, trade, etc);
Sustenance-driven collective stake in protection and regeneration of NRB	Disintegration of collective stake in NRB;
Close proximity and access-based functional knowledge/ understanding of limitation and usability of NRB	Marginalization of traditional knowledge, and imposition of generalized solutions from above;
Local control of local resourc/decisions; little gap between decision makers and resource users	Legal, adminstrative, fiscal measure displacing local controls/ decisions; wider gap between decision makers and local resource users.
Bottom line: Collective stake in NRB supported by local control and functional knowledge of NRB	*Bottom line:* Loss of collective stake and local control over NRB; resource users respond in a 'reactive' mode
C. *Social responses to (B)*	
Evolution, adoption of resource use systems and folk technologies promoting diversification, resource	Extension of exernally evolved, generalized technological/ institutional interventions; disregarding local concerns/

protection, regeneration, recycling, etc.

Resource use/demand rationing measures

Formal/informal institutional mechanism/group action to enforce the above

Bottom line: Effective social adaptation to NRB

D. *Consequences*

Nature-friendly management systems

Evolved and enforced by local communities

Facilitated by close functional knowledge and community control over local resources and local affairs

Bottom line: 'Resource-protective/regenerative' social system–ecosystem links

experiences and traditional arrangements;

Emphasis on supply side issues ignoring management of demand pressure;

Formal, rarely enforced measure.

Bottom line: NR over-extracted as open access resources.

Over-extractive resource use systems, driven by uncontrolled demands

Externally conceived, ineffective and un-enforceable interventions for protection of NRB

Little investment and technology input in NRB

Bottom line: Rapid degradation of fragile NRB; 'nature pleads not guilty'

Source: Adapted from Jodha 1995a.

users' collective experiences and knowledge of resource base, due to close proximity. Despite the presence of some inegalitarian elements, these collective arrangements worked because despite differences in individual group needs, the commonness of the source of supplies helped in integrating the individual stakes into a collective stake in the natural resource base (Sanwal 1989; see also Chapters 5 and 9).

(d) The regulatory measures and collective efforts also extended to different demand-side aspects of resource use. Collective sharing arrangements during the scarcity and crisis, management of demand pressure in general through migration and restrictions on size and composition of animal holding, etc. were quite common (Prakash 1997).

To sum up, the foundations of the traditional systems of natural resource management in the fragile areas included:

(a) The community's sustenance-driven collective or integrated stake in the health and productivity of natural resource base.

(b) Physical proximity- and practical experience-based knowledge and understanding of natural resource base, as a basis for evolving technical and institutional measures to prevent over-extractive resource use.

(c) Local control over local resources, and adherence to social sanctions empowered the community to protect and enhance community stake in its natural resources, and enforce measures which helped in balancing supply and demand aspects of resource use in the community context.

The above arrangement significantly helped in preventing the operation of P-ERD link in the past. However, as Chart 10.2 also shows, these arrangements eroded following the changes which (except population growth) were initiated from the outside. The critical and common element of these changes has been the conception, design and implementation of external interventions at the grass-roots level without sufficient understanding of the ground realities, local communities' concerns, capabilities and knowledge. These interventions in their respective ways created circumstances and perverse incentives which finally led to.

(a) disintegration of community stakes in the natural resources;

(b) disempowerment of the communities to manage the grassroots-level problems, including natural resource protection; and

(c) marginalization of local knowledge systems and institutional arrangements which helped in enforcing NRB protection.

Chart 10.1 focuses on these aspects by indicating the provisions which went against the above traditional arrangement and marginalized them without providing effective substitutes.

To elaborate further, in the first place enhanced physical, administrative and market integration of traditionally less accessible, marginal areas in the mainstream systems reduced the crucial (if not total) dependence of local communities on local NRB. Integration brought several gains to these areas, including external linkage-based diversification of sources of sustenance. But it had some backlash effects in terms of: (i) dilution or disintegration of collective community stake in its NRB; (ii) disregard and erosion of the traditional arrangements which in the past helped in the protection and regulated use of NRB; a consequence of imposition of several externally conceived and designed technical institutional interventions with little understanding and sensitivity to grassroots-level realities; (iii) depriving the local communities of their role and responsibility in managing local resources and local affairs; this happening through (a) introduction of largely outward-looking and politically oriented formal institutions such as village panchayats; (b) empowering of government revenue officials or forest officials as custodians of community NRB; (c) replacement of locally evolved institutional arrangements and customary provisions by legal and administrative arrangement evolved at higher level; and (d) distortion of community incentive system by patronage, subsidies, relief, etc. that added to the pressure on local resources without regenerating them.

The point of concern here is not the integration and its benefits, but its process. The process of integration of marginal/fragile areas including designing and implementation of intervention for them is guided by the perspectives of the mainstream system. Since the latter has largely been incapable of seeing and understanding the positive dimensions of poor and marginal areas/communities, the latter could not get enough consideration in the process of integration and transformation of fragile resource zones. For instance, the low accessibility (or poor mobility) and internal heterogeneity of these areas called for a high degree of decentralization and diversification. The interventions, without caring how things had been done in the past, focused on centralization and intensification/standardization (Sanwal 1989; Jodha 1995a). This applied to virtually every sector, including

Chart 10.2. Approaches and Constraints to Revival of Key Elements of Traditional Resource Use Systems in the Present Context

Community Stake in Local Natural Resources	Local Control over Local Natural Resources	Recognition and Use of Resource Users' Perspectives and Traditional Knowledge System
Constraints Formal legal, administrative, fiscal controls/restrictions creating a range of perverse incentives; reactive mode of community behaviour as individuals Highly depleted status of NRB creating no hope and incentive to have a stake in it More diverse and differentiated communities with different, individual rather than group-based views on community resources	*Constraints* State's in-built resistance to self-disempowerment through passing decision-making power to local communities; focus on 'proxy arrangements' e.g. village panchayats Faction-ridden rural communities; driven by diverse signals and concerns NGOs as key change facilitating agents, often governed by own perspectives, concerns	Top-down interventions with a mix of 'arrogance, ignorance and insensitivity' towards local perspectives and traditional knowledge systems Focus on (old context-specific) forms of traditional practices rather than their rational for use in the current context Rapid disappearance and invisibility of indigenous knowledge
Possible remedial approaches Genuine local autonomy for local resource management (see the next column for constraints to this); legal framework and support system for NR user groups Resource protection, investment and use of new technologies for regeneration/	Genuine decentralization, decision-making powers and resources to communities; raising the latter's capacities to respond to the above (with the help of NGOs) Rebuilding 'social capital', mobilization and participatory methods	*Possible remedial approaches* Promotion of bottom-up approaches to resource management strategies, using participatory methods and NGO help Focused efforts to identify present-day functional substitutes of traditional measures for resource management

high productivity of NRB (using experiences of successful initiatives)
Collective stake through planned 'diversification' and 'shareholding' system in natural resource development and gains (using experiences of successful initiatives)

using NGO input; focus on diversified, high-value products from rehabilitated NRB (using successful experiences)
Required changes in NGO approaches/perspectives by introspection; involving small local groups, and unlabelled agencies

R&D to incorporate rational of traditional knowledge system (using experiences of successful initiatives)

Source: Adapted from Jodha 1995a, 1997.

management of local natural resources. Hence, the loss of traditional arrangements for protection and regeneration of NRB is largely a consequence of the specific approach to integration and development of fragile resource areas.

IS THERE A SOLUTION?

The preceding discussion has tried to identify some key factors and circumstances responsible for NRB protection in the past and their disregard and decline leading to resource degradation in the changed context of today.

In view of the recurrent failures and ineffectiveness of interventions directed at stopping human-induced degradation of NRB, one may be tempted to look for some lead from the traditional arrangements. Before venturing in this direction, it should be clearly stated that pleading for revival of traditional arrangements for natural resource management may be an exercise in futility, because most of the objective circumstances associated with them have completely changed to permit their revival and ensure their effectiveness in the present-day context. For instanc_: (a) market penetration and changes in the attitude of village communities have promoted the values and approaches which put very low premium on collective strategies; (b) population growth, rise in factionalism, and increased economic differentiation have made it difficult to evolve and maintain a community stake in natural resources; (c) depletion of NRB and depletion of the culture of group action tend to reinforce each other in accentuating community indifference to rehabilitation of NRB for collective gains; (e) the legal, administrative and fiscal mechanisms (despite lip-service for the opposite) have a strong tendency for centralization and application of uniform, generalized solutions, ignoring the diversities at the grassroots level.

The plea here is not for revival of traditional arrangements as they were. Instead, our focus is on a search for functional substitutes of the traditional arrangements, which can fit with the present-day circumstances.

THE THREE PILLARS OF TRADITIONAL SYSTEMS

Three elements of traditional resource management/usage systems, which in the past played a crucial role in preventing human-induced degradation of natural resources in the fragile resource zones, along with the objective circumstances that promoted and strengthened them were:

(a) strong community stake in its NRB, facilitated by the community's near-total dependence on local NRB;
(b) local control over local resources resulting from isolation and inaccessibility which induced a degree of autonomy;
(c) resource users' and decision-makers' functional knowledge of limitations and usability of their diverse NRB resulting from people's close physical proximity and access to resources.

The incorporation of the three elements (i.e. community stake, local control and functional knowledge of NRB), into the present resource use systems may help in rehabilitation and conservation of NRB, and therefore should be promoted. But revival of their historically associated objective circumstances (e.g. exclusive and almost total dependence on local resources, semi-closed communities, physical proximity) is neither possible nor desirable. The challenge lies in creating a present-day functional substitute of the past circumstances, which can promote the three key elements (community stake, etc.) and induce communities to protect and regenerate their natural resources while using them.

In the following discussion a few possible ways are indicated to attempt it. The discussion is focused on the constraints to revival of the three elements and possible remedial measures to handle them (Chart 10.2). The suggested possibilities could just be loud thinking on the subject but they are supported by small and scattered evidence based on successful experiences.

(A) Reviving Community Stake in NRB

Community stake in the local NRB is central for the protection of natural resources, but in the present-day context there are more circumstances discouraging this than supporting it.

(i) *External Controls and Perverse Incentive Systems:* Local communities mostly respond or simply adjust to external interventions and impositions, i.e. government laws and regulations, rather than control or plan their approach to NRB. The whole incentive structure—permitting privatization of community resources, illegal extraction with little penalty, priority to political patronage and unrealistically low or little pricing of high-value natural resources products—is designed and operated against community involvement in resource-protective and regenerative efforts. Possible solutions to some of these problems lie in effective transfer of local resources to local control and genuine involvement of local communities in the management of NRB.

(ii) *Extremely Depleted NRB Not Worth a Stake:* In the context of present biophysical (and economic) status of NRB, local control over local resources may not induce positive response of the community. Natural resources in many areas are depleted to a level which does not inspire much hope, let alone the community's group action for their management. Nevertheless, if one goes by the field evidence on rapid regeneration of natural resources with some protection in many parts of India (Poffenberger and McGean 1996; Hazra et al. 1996) and more impressive results in Nepal, the doubts on rehabilitation and growth of biophysical productivity of NRB may prove misplaced. At the same time, the need for investment and technological input (which is not even a fraction of the efforts currently devoted to crop lands) cannot be overstated. Additionally, in the changed economic context, the focus on NRB as an exclusive source of biomass may not help. To induce and encourage active and effective involvement of communities in NRB protection, emphasis should be placed on harnessing high-value products (herbs, flowers, seeds, etc.) from the community's natural resources. Evidence shows that people care more about the more productive unit than unproductive unit of the same type of community resource in the same village (see Chapter 5).

(iii) *Increased Economic Differentiation and Diversity of Interests:* The technical and economic issues relating to productivity and rehabilitation of NRB discussed above are much simpler to address. The bigger problem relates to the reconciliation of interests of diverse groups in the villages, without which a community's collective stake in NRB is impossible. Internal heterogeneity and inequities are not new in the Indian (South Asian) villages. However, following the already mentioned changes (Chart 10.1), decline of the culture of group action, increased economic differentiation and socio-political factionalism, the differences and divisions in the rural communities have greatly increased.

Also, the traditional circumstances facilitating informal inter-group bargaining and reconciliation (Leach et al. 1997) have vanished. For instance, in place of local NRB as common source of sustenance, now there prevail multiple and diverse sources of sustenance (of internal and external origin) for the village economy; the long lead time available for internal adaptations and bargaining by action, is no more available; socio-political contexts for different groups at times also fall outside the boundaries of the local community's influence. All these factors will potentially obstruct the evolution or revival of the community's collective stake in NRB.

Remedial Measures: Approaches to Promote General Action. Most of the problems indicated above are of institutional nature. We shall address them later while discussing the issues related to local control over local resources. At this stage we shall focus on identifying the 'technical (?) basis' of common stake in NRB for a diverse community.

First, promotion of genuine functional local autonomy over local resources is the key step. An important step in this context is to provide an institutional and operational framework for establishment of the NRB-user groups, as already successfully attempted in countries like Nepal. This approach has strengthened collective community stake in natural resources. A somewhat similar but less dramatic experience one finds in areas with joint forest management in India (Poffenberger et al. 1996).

A major problem with this approach emerges when the NRB legally belongs to the whole village, but all villagers do not subscribe to similar use or the same product (e.g. fodder, timber) from community resource. For instance, the richer group may want growth of timber or commercially more valuable product (after long waiting), while the poor may prefer more of biomass for current use.

Babhar grass (in north Indian villages) is an illustration. It is good fodder when young but fetches a high price from the paper mill when sold mature. The options are tailor-made for a conflict of interests between the poor and the rich. Similarly, water harvesting under integrated development of watershed (which belongs to the whole group) may not help those who do not have land to irrigate.

In cases like the above, a planned diversification of resource use, including processing and marketing activities or introduction of a system of share holding for service or products sharing on an equal basis can be attempted. This has already been done in Shukhumajari watershed development project (Sarin 1996), which has helped in the establishment of return-based common concern for NRB.

If complemented with investment and technology as well as promotion of high-value options discussed earlier, the above approach can help in reviving the community stake in its natural resources. An external input through involvement of NGOs can further facilitate the process. The key lies in the identification of 'product' which different groups can share.

(B) Local Control over Local Natural Resources

Traditionally, the mainstream decision-makers permitted greater local autonomy to communities in fragile resource zones. This was more due to default (i.e. their inaccessibility-imposed ignorance and indifference) rather than a conscious decision. With the increased physical and administrative integration of fragile, remote, marginal areas with the mainstream political–economic systems, most of the local natural resources belonging to the communities were taken over by the state either through formal law or through disregard of customary laws and practices (Poffenberger et al. 1996; Jodha 1996b). The consequent lack of local control over local resources prevents local decisions and action for protection and regulated use of natural resources. The importance of changing this situation can hardly be overstated.

Genuine and effective restoration of local control over local natural resources is, however, faced with several constraints from both the state and the local communities.

(i) *State's Resistance to Self-disempower:* Despite all talk of decentralization and power to people, etc., when it comes to the control of a property or productive resources, the state operating through its sectoral bureaucracy always tries to avoid the issues. Either it tries half-hearted compromises such as under the Joint Forest Management (JFM) in India, where the community is involved in protecting resources and limited sharing of specific products (e.g. timber) plus use of products which the state finds difficult to use (e.g. fodder, minor forest products, etc.).

Another approach adopted by the state is to create formal institutions such as village panchayats, with all legal powers and provisions as decided by the decision-makers at the top. Most such bodies are small-scale political bodies with very little concern and involvement in NR management; except when relief and subsidies could be mobilized by showing the extent of community resources the village has (see Chapter 5). These bodies (despite the recent focus on genuine decentralization) may not be a substitute for 'user groups', as their goals are too diversified, and NRB constitutes a small component therein. The difference between village commons managed through village elders and those managed by elected panchayat makes this aspect clear (Brara 1987). The solution to these constraints lies in genuine decentralization, empowering the natural-resource-user groups, and social mobilization efforts with the help of NGOs, etc. At the same time, these measures too have certain constraints, as discussed below.

(ii) *Faction-ridden, Differentiated Rural Communities:* As already alluded to, the present status of most village communities, characterized by factionalism, high dependence on government patronage, and completely eroded culture of group action and sharing systems, does not equip them strongly for accepting and effectively implementing the responsibilities associated with transfer of local resources to local control. In fact some elements of such transfer already form part of the panchayat system, with little visible impacts so far.

The solution lies in a gradual rebuilding of what is described as social capital, implying culture and mechanisms promoting trust, sharing and group action. Though a crucial requirement of the local-level natural resource management, this institutional change cannot happen through formal, legal measures only. This is a task of social awareness generation and mobilization, where grassroots-level voluntary agencies (NGOs), complemented by genuine encouragement by the state can help. A number of participatory rural development initiatives are already in place (Krishna et al. 1997; Zazueta 1995). Learning from the existing success stories in this field and efforts to replicate them could be an effective approach to equip the rural society to manage local resources and build a collective stake in their NRB.

(C) Use of Local Perspective and Traditional Knowledge Systems

Even when the advocacy on natural resource management is conceived in a national or global context, in most cases its practical context relates to the local or micro levels. Hence, unless the perceived and projected approaches to natural resource management are sensitive to the local-level perception and problems, their success may be limited. However, the local community perceptions, specially the traditional knowledge and experiences are usually bypassed while planning and initiating interventions for local areas and communities.

A key constraint against changing this approach is the attitude (involving some arrogance and insensitivity) of the planners of the top-down approaches to solve local-level problems, including degradation of community NRB. Another reason for bypassing traditional knowledge systems is its general non-availability in very articulated form on the one hand and the focus on the form rather than rationale of the traditional practices, on the part of technocrat decision-makers on the other. Since the forms of traditional practices had been context-specific (e.g. land-extensive farming practices worked well under low population pressure, or total dependence on local NRB helped in building community stake in NRB), they became unfeasible or ineffective with the changed context. The decision-makers, instead of evolving alternative forms, have discarded both the rationale and forms of traditional practices.

Remedial Measures: The formulation of the above constraints itself suggested some remedial approaches. Accordingly, the focus on bottom-up approach to natural resource management; sensitization of decision-makers to the local community's perceptions through partici-patory approaches; identification and incorporation of rationale of traditional practices into new technological and institutional measures planned for natural resources, should be encouraged. Some ongoing initiatives supported by NGOs are already using these approaches.

NGOs AS CRUTCHES

Three key actors involved in the proposed approaches to rehabilitate and sustainably use the community natural resources are the state, the village communities and NGOs. The factors constraining the role of the first two and possible remedial approaches have already been discussed. NGOs are a new actor on the scene. Many of them are mandated to perform or facilitate the changes advocated in this essay.

NGOs, however, are a mixed category, and just being an NGO by itself is not a sufficient condition for their involvement and effective-ness in the tasks outlined here. A discriminating approach in choosing the agencies is called for.

Secondly, by background most of the NGO workers are urban groups even when they operate in rural areas with genuine concern. Consequently, their orientation and perspectives may be at variance from the genuine rural (unarticulated) perspectives on the problems. In such cases, NGOs would be only marginally different from the planners of top-down interventions. Even genuinely grassroots-level rural voluntary groups in the process of their upward graduation get sucked into the mainstream (urban-centred) NGOs. Possible solutions to this problem may include federating the local voluntary groups, non-imposition of predetermined perspectives of NGOs but to evolve through participatory approaches solutions to rural problems. A number of NGOs already do this, but there are no objective yardsticks to separate such NGOs from others on a *priori* basis. NGOs themselves have not been able to evolve any quality control devices for this purpose.

The third aspect, quite related to the above issues, concerns the role of NGOs as mediating agency between the state and rural community. In the process (as apprehended), NGOs may not only focus on selling their own perspectives and approaches but tend to build their own space and indispensability. That fits well with the state which neither understands rural communities well enough nor can deliver promised goods and services. NGOs fill this gap comfortably. A possible solution lies in the NGOs' own determined effort to reduce communities' dependency on them and create local voluntary groups while implementing the interventions, and not to stick to the same place longer than needed.

Fourth, the tendency of many NGOs to stick to their success stories and use them both for greater limelight and increased funding from governments and donors compromises their image as task-oriented voluntary groups. A possible solution will be the same as indicated in the preceding paragraph.

Field experience also shows that several government agencies, academic institutions and unlabelled small voluntary groups have helped in resolving grassroots-level problems, both by field action and influencing policy programme. Their strong complementarity with the NGO efforts should be further promoted and harnessed (Zazueta 1995; Krishna et al. 1997). This probably can help in replicating and up-scaling the success stories of community natural resource management scattered in different pockets.

11

Globalization: Repercussions for Fragile Ecosystems and Society

In the preceding chapters we have frequently referred to the fact that people's adaptation strategies, through diversified production activities, traditional resource management systems, collective stakes and group action, are gradually declining under demographic, economic, institutional and technological pressure of changes, in both mountains and dry tropical regions. Their environmental and socio-economic consequences were also indicated.

Compared with the past, the fragile ecosystems and their communities are likely to face still greater pressure of changes in the twenty-first century. The key developments, with both positive and negative consequences, would include changes in the roles of the state, market and civil society; rapid growth and application of information and communication technology, continued population growth and rise in external demand for fragile-zone products; increased environmental degradation and multi-level concerns over it, and the rapid process of economic liberalization or globalization with enhanced primacy of market forces in development discourse and design. In this chapter we elaborate primarily on the last category of change and its links with other changes, as this is expected to be potentially the most powerful development affecting all countries, regions and ecosystems. Furthermore, our focus is on potentially negative consequences of globalization, without minimizing its possible positive contributions. We may also alert the reader that in terms of evidence and analysis this essay does not match the contents of preceding chapters, as work on this subject has been initiated only recently. It is largely based on extrapolation of the past understanding of the fragile-resource situation *vis-à-vis* the rest of the world and some observations on the emerging scenario. The purpose of the essay is to alert all those interested in fragile ecosystems as well as their environmental and socio-economic problems about the likely negative consequences of strong market-driven interventions in these areas and induce thinking, research and practical steps to minimize such consequences.

WHAT IS GLOBALIZATION?

Put simply, the globalization process implies adoption of market-friendly economic policies and programmes specifically directed to liberalization of trade and exchange policies, reorienting development and investment priorities and restructuring of rules and provisions guiding economic transactions as well as roles of different actors in the process, as dictated by the pressures and incentives generated by global economic forces and their legal, fiscal and institutional instruments (UNDP 1999). Its key implication relevant to present discussion is the fact of according primacy to global perspectives and external concerns even while dealing with local problems, and in the process disregarding local perceptions and practices. The mechanisms through which global perspectives could be imposed at micro-level (or in fragile-ecosystems context) are commodity trade and associated resource use as well as production patterns, deregulation of existing economic provisions and their enforcement measures, curtailment of welfare and promotional support for the needy, promotion of preferred technologies and support systems through a range of investment, tax and price incentives, etc., as induced by market requirements, which in turn are insensitive to both environmental and social concerns (Norgaard 1999).

The presumed virtues of globalization include greater gains of free flow of resources and products ensuring more efficiency as well as greater growth of wealth and welfare at the global level and assigning of the development and distribution business to the forces of market, which through incentive-driven transactions can perform this business more efficiently (World Bank 1999). Besides the hidden agenda of contemporary international political economy, the advocacy of globalization is also characterized by a number of questionable assumptions behind it (South Centre 1996). They become clearer when the process of globalization is viewed in the micro-level context. We attempt this with reference to fragile ecosystems and their communities.

Broadly, the process of globalization creates circumstances which are beyond the control of communities in fragile ecosystems. It is governed by driving forces insensitive to the concerns of fragile ecosystems. The process is so rapid and overpowering that affected communities have neither sufficient lead time nor required capacity to adapt. As a final consequence, globalization tends to accentuate the process of exclusion of local communities from both their resource base as well as the pace and pattern of rapid economic and social change in fragile zones. In particular the exclusion process is dominated by resource degradation and marginalization of well-adapted production options and practices which helped in environmental sustainability and livelihood security of people in the fragile areas. More specific and interrelated contexts for understanding potential repercussions of the rapid globalization process on fragile ecosystems and their dependent populations are elaborated below.

POTENTIAL IMPACTS AND PROCESSES

A. Visible Incompatibilities Between Driving Forces of Globalization and Imperatives of Specific Conditions of Fragile Ecosystems

The globalization process is driven by market forces which (guided by short-term profitability and external demand) promote selectivity and narrow specialization in the choice of production activities, encourage indiscriminate resource-use intensification, and over-extraction of niche opportunities/resources with little concern for their environmental and socio-economic consequences. These orientations are directly in conflict with the imperatives of specific conditions of the fragile areas rooted in their high degree of fragility, marginality, diversity, specific niche, etc. These specific features create objective circumstances which favour diversification of resource use and production activities, balancing of intensive and extensive use of land resources as well as that of production and protection needs. Despite the said incompatibilities, the market forces and associated incentives, facilities and pressures generated by globalization are likely to enhance indiscriminate resource-use intensification both on farm/community as well as niche-resource focused corporate firm level. The negative environmental impacts will be their first consequence. An equally important implication would consist of disregard and erosion of resource-use systems and production practices evolved over generations to have sustainable environment and livelihood security in the fragile ecosystems, particularly in mountain areas. Some evidence of the above process at farm level is already visible through (i) exclusive focus on oilseed crops and other high-value crops leading to mining of ground water in parts of the dry zone, and (ii) selected horticulture crops in hills. The environmental and productivity impacts of monoculture or reduced diversification are also increasingly felt. Over-extraction of resources (timber, mineral, hydropower, herbs in mountain areas) with its negative side-effects is also well recognized (see Chart 11.1).

B. Possibility of Accentuating the Negative Impacts of Past Intervention

It may sound strange, but as far as fragile ecosystems (mountains and dry tropics) are concerned, the past public-sector-determined development interventions and the new market-driven processes under globalization share a number of common elements. They include extension of externally conceived and designed, very much standard-ized and highly top-down interventions to fragile areas with little concern for biophysical and social circumstances; indiscriminate resource-use intensification with little concern for fragility and diversity; over-extraction of niche resources to meet external demands, imposition of external perspectives, institutions and technologies, marginalizing the traditional well-adapted systems (see Chapters 3, 4, and 9). These elements had been the source of negative side-effects of development interventions in the fragile areas. The examples can be located in the past production programmes (e.g. grow more food or high-yielding-variety-based sole cropping systems in agriculture), development programmes and welfare activities, (e.g. public distribution system or cooperative institutions); integrated rural development schemes or employment guarantee schemes, etc., where unmodified activities were extended from prime agricultural areas to the fragile zones, without much attention to their specific circumstances. The intended resource-use intensification for high production led to rapid resource degradation such as rapid depletion of ground water in dry areas, landslides and soil erosion in mountains. This weakened the people's traditional practices, activities, as well as group action for resource management, including common property resources (see Chapters 4, 5 and 9). The globalization process, governed by external market forces (and being much less sensitive to local circumstances), is likely to accentuate the above trends. Gradually the weakened state, yielding to the incentives and pressure from the

Chart 11.1. Potential Sources of Adverse Repercussions of Globalization for Fragile Ecosystem and Communities

Potential Sources	Elaboration/Examples
A. Visible incompatibilities between driving forces of globalization and imperatives of specific features of fragile ecosystems (fragility, diversity, etc.)	*Incompatibilities* Market-driven selectivity, resource-use intensification and over-exploitation of niche induced by uncontrolled external demand versus fragility-marginality–induced need for balancing of intensive and extensive resource uses; diversification of production systems, niche harnessing in response to diversity of resources *Consequence:* Environmental resource degradation; loss of local resource-centred, diversified livelihood security options; increased external dependence
B. Accentuation of negative side-effects of past development interventions under globalization due to their common elements (approaches, priorities, etc.) with adverse effects on fragile areas	*Common elements between past public interventions and market-driven globalization* • Externally conceived, top-down, generalized initiatives (priorities, programmes, investment norms) with little concern for local circumstances and perspectives, and involvement of local communities • Indiscriminate intensification at the cost of diversification of resource use, production systems and livelihood patterns causing resource degradation (e.g., deforestation, land slides in mountains; mining of ground water to irrigate high water consuming crops in dry areas) • General indifference to fragile areas/people excepting the high potential pockets creating a dual economy/society; over-extraction of niche opportunities (timber, mineral, hydro-power, tourism) in response to external (mainstream economy) needs, with very limited local development *Consequence:* Environmental degradation and marginalization of local resource-use systems, practices, and knowledge, etc. likely to be enhanced due to insensitivity of market to these changes and gradually weakened public sector

C. Globalization promoting erosion of provisions and practices imparting protection/resilience to fragile areas/people (including disinvestment in welfare activities)	*Marginalization Process* • Traditional adaptation strategies based on diversification, local resource regeneration, collective sharing, recycling, etc., likely to be discarded by new market-driven incentives and approaches to production, resource management activities • Shrinkages of public sector and welfare activities (including subsidies against environmental handicaps, etc.) depriving areas/people from investment and support facilities (except where externally exploitable niche opportunities exist) *Consequence:* Likely further marginalization of the bulk of the fragile areas and people
D. Loss of local resource access and niche-opportunities through the emerging 'exclusion process'	*Exclusion Process* Niche resources/products/services with their comparative advantage: • For mountains: timber, hydropower, herbs, off-season vegetable, minerals, eco-tourism • For dry tropics: drought resistant biodiversity, specific spices, oil seeds, animal products, minerals, eco/cultural-tourism Their likely loss under globalization through: • Market-driven over extraction/depletion due to uncontrolled external demand • Focus on selective 'niche', discarding diversity of niche, their traditional usage systems, regenerative practices • Transfer of 'niche' to mainstream prime areas through market-driven incentives, green house technologies, infrastructure and facilities (e.g. honey, mushrooms, flowers produced cheaper and more in green house complexes in the Punjab plains compared to naturally better suited Himachal Pradesh, OGL making Himachal Apple uncompetitive) • Acquisition and control of access to physical resources: forest, waterfalls, biodiversity park, tourist attractions by private firms through purchase or auction by government, depriving local's access, destroying customary rights and damaging livelihood security systems *Consequence:* Loss of comparative advantages to fragile areas or access to such gains for local communities

(Continued)

Chart 11.1 continued

Potential Sources	Elaboration/Examples
E. Adapting to globalization process possible approaches to loss minimization	• Sharing gains of globalization through partnership in primary and value adding activities promoted through market; building of technical and organizational capacities using NGOs and other agencies including market agencies to promote the above • Promotion of local ancillary units (run by locals) to feed into final transactions promoted by globalization; this needs institutional and technical infrastrutre and capacity building • Provision for proper valuation of fragile areas resources and compensation for their protection, management by fragile areas people for use by external agencies • Enhance sensitivity of market-driven initiatives to environment and local concern to be enforced by international community (e.g., WTO) and national government • All the above steps need local social mobilization, knowledge generation and advocacy movements, and appropriate policy-programmes *Consequence:* If above steps are followed, there are chances of influencing the globalization process and reducing its negative repercussion for fragile areas/people; help in adaptation to globalization

Source: Jodha, 1999.

globalization process, would find it increasingly difficult to act against the accentuation process.

Another feature of past interventions was the coexistence of policy-makers' general indifference towards fragile areas and their strong focus on niche opportunities of fragile areas, which could be exploited for the mainstream economy. In dry areas the niche opportunities included oilseeds, pulses, specific spices and related products and animal products. For mountains there are many more, and more significant, niche resources such as timber, hydropower, horticulture and herbs. Minerals and tourism offer attractive opportunities both in mountains and dry areas. Policy-makers and planners (quite similar to market-driven entities) have focused disproportionately higher efforts and investment to extract these opportunities. In the process while local populations benefited a little, the bulk of the gains went to the mainstream economy outside these regions. The environmental and long-term sustainability costs of such resource extractions are huge and rarely compensated. In the case of mountains, due to unequal high-land–low-land links the uncompensated flows of resources and products to low lands is extremely high (Banskota and Sharma 1999).

The purpose of the above elaboration is to indicate that past interventions due to their insensitivity to specificities of fragile resource zone had several negative side-effects. Since the market-driven interventions under globalization process are likely to be more indifferent to fragile area features, the negative impacts stated above would accentuate further. This is all the more likely because not only market forces focus on selectivity rather than diversification, but the technologies, support systems, organizational and decision-making culture associated with the globalization process are far removed from the realities of fragile resource zones.

EROSION OF PRACTICES AND PROVISIONS IMPARTING RESILIENCE AND PROTECTION (INCLUDING WELFARE PROGRAMMES)

Two broad categories of provisions and practices have helped fragile area people in the past. First, the communities in mountains and dry tropics have evolved their own adaptation strategies to ensure both protection and use of fragile and marginal resources as well as security of their livelihood (see Chapters 3 and 9). These strategies are manifested through diversified and flexible resource use, resource recycling, risk sharing arrangement, etc. Despite their decline in recent decades, these practices are still an important part of their economic and social transactions. To this one may add the gains from local harnessing and exchange of niche products with comparative advantage to fragile areas.

Second, despite their limitations, public policies, through welfare programmes and subsidized development interventions, such as promotion of pasture development, water harvesting, and R and D and micro-enterprise activities, have been helping the fragile area people to compensate for natural and other handicaps faced by them. The public sector plays a crucial role in these activities.

These protective provisions and practices are likely to decline due to pressures generated by globalization. In consequence, the traditional practices, notwithstanding their continued rationale and utility, are likely to be disregarded and marginalized by market-driven processes under globalization. We have already alluded to the traditional practices and arrangements which will have serious backlash from the new short-term profitability-centred production and resource management systems driven by tradability and external perspectives. There is a strong possibility of the emergence of a dual system consisting of rich and resourceful groups/pockets participating in the change process and the bulk of the poor left with limited options. This is already visible through emerging gaps between the progressive and transformed pockets participating in market processes and the bulk of fragile areas still remaining out.

Similarly, with rapid shrinkage of the public sector and role of the state, as well as changed efficiency and productivity norms for resource allocation under the strong 'market dominated regimes', both welfare- and subsidy-supported development programmes are likely to be de-emphasized (Jodha 1996a). The consequent disinvestment in welfare and protective programmes is already emphasized under structural adjustment plans (Roy 1997). Yet another major negative consequence of globalization would take the form of fragile zone communities losing their niche resources and opportunities. This forms a part of exclusion process, as elaborated below.

Loss of Niche and Access to Opportunities: An Emerging 'Exclusion Process'

Both mountains and (to a lesser extent) dry tropics are endowed with unique environmental and resource characteristics, which have potential for products and services with comparative advantage to these areas. As already mentioned timber, hydropower, off-season vegetables, seed production, valuable herbs, minerals and eco-tourism, etc., constitute niche of mountains. Drought-resistant biodiversity, specific spices, oilseeds, hardy legumes, animal products, minerals and eco-tourism are unique to dry areas as well. Under the market-driven compulsions and facilities these areas may face loss of their niche. The process is likely to include the following.

First, the survival and sustainable use of niche resources is closely associated with their protection while using them and the links between diversified resource-based activities. Both these conditions may not be satisfied in the face of external market-driven pressures and incentives for selective over-exploitation and indiscriminate resource-use intensification. The adverse environmental and socio-economic consequences of such changes have already been alluded to.

Second, globalization process would bring in new sets of incentives, technologies, infrastructure and support systems, which in response to high demand and profitability would facilitate creation of man-made facilities for production of items outside fragile areas, in which the latter hitherto had comparative advantage. Already, one comes across several developments of this nature. For example, products such as honey, mushroom, flowers, herbs, off-season vegetables and quality crop seed, hitherto produced mainly by mountain areas such as Himachal Pradesh, are now produced much cheaper and in a larger quantity under the massive greenhouse facilities in the plains of the Punjab. There is yet another development encouraged by the trade policies, which marginalizes the niche opportunities of fragile areas, by way of substituting fragile areas' niche products by imports (due to OGL) such as hill apples replaced by imports from abroad; local oilseed products and milk products from dry areas replaced with imported milk products, soyabean and palm oil, etc.

A related trend, which tends to restrict fragile area populations from harnessing their niche opportunities, is manifested by the provisions of global conventions/treaties such as those relating to biodiversity conservation, desertification, forest policies and climate change, etc., which focus on the protection of 'global commons', without sufficient concern and understanding of fragile areas and their inhabitants directly dependent on the 'global commons' (Jodha 1997; Zerner 1999).

The possibilities of the above losses being compensated through fragile areas/communities sharing the gains from globalization are restricted because of their inability to participate in the change process.

The 'exclusion process' indicated by the above possibilities is further accentuated by the alienation of the local communities from their niche resources and associated niche opportunities. Accordingly, in the situations where due to physical or economic inseparability of niche from fragile ecosystems, the marginalization of niche opportunities of fragile zones is not possible though the aforementioned processes, a different pattern of depriving the local communities from their niche resources is emerging. This involves external agencies (MNCs and national firms, etc.) increasingly acquiring ownership, or exclusive access and usage rights to landscapes and specific resources in fragile zone. Disregarding the customary right and local control and access to such resources and products, large-scale areas are given by the state to private companies in the name of resource development and product harnessing. Auctioning or leasing of so-called wastelands, development of herbal farms, leasing of areas for mining, right to water flows for hydropower, forest for timber, enclosures for parks and biodiversity, prime spots for tourism resorts (and private dwellings for the rich from outside) are some of the examples of changing the ownership and access to resources in the fragile areas, more so in mountain areas. These developments depriving the local communities from their own resources, are complemented by the already mentioned global initiatives manifested by global treaties and conventions, where enlightened policy-makers rather than market forces play the key role.

POSSIBLE WAYS TO INFLUENCE AND ADAPT TO GLOBALIZATION PROCESS

The overall scenario described above portrays a rather bleak future for the fragile ecosystems and their communities. At the same time, in view of the unfolding realities at national and international levels, it is not possible to wish away the process of globalization. The best option lies in eliminating or minimizing its negative repercussions for fragile areas.

The solution to the potential problems stated above lies in influencing and modifying the said processes and adapting to the changes led by globalization. The approach to achieve these objectives may include a context-specific mix of multiple steps. The specific focus of the steps could be on minimizing economic losses, preventing exclusion possibilities, ensuring local participation in the resource harnessing decision, creating compensatory mechanisms for environmental services offered by fragile areas and their people to the rest of society and the economy, etc. Some loud thinking on these issues is presented below.

To begin with, one should focus on the mechanism which can help in fragile areas people sharing the gains from globalization through their engagement in primary and value-adding activities based on fragile-area opportunities promoted by globalization. This implies their participation in the market-driven process of change. To facilitate their participation, the local people have to be equipped in terms of requisite skills, capacities, etc. This in turn requires social mobilization and technical as well as organizational/management training. A number of NGOs in specific context are already attempting this in scattered locations and activities. The private firms entering fragile zones could be involved in the process by demonstrating the utility of local participation in market-driven initiatives. The local perceptions could prove immensely useful in dealing with the environmental implication of the new resource-intensive ventures likely to be encouraged by market forces in the fragile ecosystems.

One of the most effective ways of ensuring local participation in external initiatives in fragile ecosystems is to associate local communities through ancillary activities to support the main production/resource harnessing ventures as has been attempted in China (Rongsen 1998). However one has to guard against the ancillary activities becoming exploitative for the local participants.

Yet another means for local communities sharing the gains of globalization is provision of adequate compensation for their losses through various forms of exclusion process discussed above. The recognition of customary rights and related practices; protection of intellectual property rights (IPR) are other issues, which can form part of compensation or basis for local partnership in market-driven initiatives.

The biggest issue requiring compensation relates to the current pattern of uncompensated flow of resources and products from fragile ecosystems (particularly mountains) to mainstream, urban economy. There is urgent need to evolve valuation procedure to assess the real worth of resources (timber, water, hydropower, environmental services, tourism, specific natural products, etc.), which are largely protected and regenerated through the resource management practices of the communities in fragile areas. Thus it is their effort and investments, which helps ensure availability of environmental resources and services to the mainstream economy. If this task is performed well and appropriate compensation measures are worked out, most of the fragile ecosystem (specially mountain areas) would not have to look for charity and subsidy from any agency. If the globalization process is made accountable for externalities and induced to fully compensate for the resources and services it uses, the loss due to shrinking public sector and disinvestment in welfare activities could be more than compensated.

The second strategic step to influence or modify the globalization process is to sensitize the market-driven decisions and actions to the environmental concerns in fragile areas. Since globalization normally follows the signals provided by market forces, the proposed sensitization is not an easy task. Yet, if the international community is serious about its concerns presented in different fora some strict limits in extracting resources and manipulating the environment will have to emerge from international agencies (e.g. WTO, UNEP, the World Bank) and national governments. However, as alluded to earlier, these agencies as promoters of global treaties and conventions affecting fragile areas, themselves are too indifferent towards the imperatives of specific features of fragile ecosystems such as fragility, marginality, diversity, and their interlinkages, etc. to influence them. However considerable knowledge-based advocacy at national and international levels will be required.

This task would need mobilization of local communities and sympathetic external voices from NGOs, environmental activists, academia, donors and sensitive government agencies. In the age of information revolution and communication technologies, linking of voices and concerns from multiple agencies and locations is not difficult. Besides, the official or semi-official agencies dealing with issues of global warming and climatic change, biodiversity conservation and desertification, etc., have considerable clout to influence the governments and international agencies. Their awareness and convictions about the consequences of economic liberalization can surely get the attention of policy-makers at the highest level.

Indicative policy challenges and choices elaborated elsewhere (Jodha 1997) to address the problems highlighted by this essay are: (i) need for sound information base and understanding of involved issues listed in the essay; (ii) evolving strategies and approaches to strengthen fragile-area niche; (iii) arresting emerging 'exclusion process'; (iv) advocacy of (and action) on 'local' responsibilities of 'global' stakeholders in fragile areas.

References

Agarwal, A. and S. Narain, 1990a. *Towards Green Villages*, Centre for Science and Environment, New Delhi.

Agarwal, B., 1986. *Cold Hearths and Barren Slopes: The Woodfuel Crisis in the Third World*, London: Zed Books.

—————, 1990. 'Social Security and the Family: Coping with Seasonality and Calamity in Rural India', *Journal of Peasant Studies* 17(3): 341–412.

Ahuja, K. and M.S. Rathore, 1987. *Goats and Goat-Keepers*, Jaipur: Printwell Publishers.

Allan, N.J.R., G.W. Knapp and C. Stadel (eds), 1988. *Human Impacts on Mountains*, New Jersey: Rowman & Littlefield.

Altieri, M.A., 1987. *Agroecology, The Scientific Basis of Alternative Agriculture*, Boulder, Colo: Westview Press.

—————, and L.C. Merrick, 1988. 'Agroecology and In-situ Conservation of Native Crop Diversity in the Third World' in E.O. Wilson and F.M. Peter (eds), *Biodiversity*, Washington, DC: National Academy Press.

Ananthram, K. and J.C. Kalla, 1988. 'Eco-Sociology of Range Resources: Spatial and Sectoral Interference in Managing the Rangeland Resource in Arid Rajasthan', paper presented at the Third International Rangeland Congress, 7–11 November, New Delhi.

Anderson, J.R., 1991. 'Aspects of Agricultural Research as Aids in Risk Management' in P.B.R. Hazell and A. Pritchard (eds), *Risk in Agriculture*, Washington, DC: The World Bank.

—————, and N.S. Jodha, 1994. 'Agricultural Research Strategy for More Eduring Productivity in Fragile Areas' in J.R. Anderson (ed.), *Agricultural Technology: Policy Issues for the International Community*, Wellingford: CAB International.

Anon, 1960. *Report of the State Land Utilisation Committee,* Government of Rajasthan, Jaipur: Government Press.

Anon, 1988. *Whither Common Lands?* Dharwad: Samaj Parivartan Samudaya (NGO).

Arnold, J.E.M and W. Stewart, 1991. *Common Property Resource Management in India: A Desk Review. Report for the Asia Environment Division and the India Agriculture Division*, Washington, DC: The World Bank.

Arnold, J.E.M. and P. Dewees (eds), 1995. 'Tree Management in Farmer Strategies: Responses to Agriculture Intensification', Oxford: Oxford University Press.

Arora, D., 1994. 'From State Regulation to People's Participation: Case of

Forest Management in India', *Economic and Political Weekly* 29 (19 March): 691–8.

Bandyopadhyay, J., 1987. 'Political Economy of Drought and Water Scarcity', *Economic and Political Weekly* 22(50).

Banskota, K., 1992. Agriculture and sustainable development. In: Asian Development Bank, *Nepal–Economic Policies for Sustainable Deveopment*, Manila: Asian Development Bank.

Banskota, M., 1989. *Hill Agriculture and the Wider Market Economy: Transformation Processes and Experience of the Bagmati Zone in Nepal*, ICIMOD Occasional Paper No. 10, Kathmandu: ICIMOD.

——————, and N.S. Jodha, 1992a. 'Mountain Agricultural Development Strategies: Comparative Perspectives from the Countries of the Hindu Kush-Himalayan region' in N.S. Jodha, M. Banskota and T. Partap (eds), *Sustainable Mountain Agriculture*, New Delhi: Oxford and IBH Publishing Co., pp. 83–114.

——————, and N.S. Jodha 1992b. 'Investment, Subsidies and Resource Transfer Dynamics: Issues for Sustainable Mountain Agriculture' in N.S. Jodha, M. Banskota and T. Partap (eds), *Sustainable Mountain Agriculture*, New Delhi: Oxford and IBH Publishing Co.

Berkes F. (ed.), 1989. *Common Property Resources: Ecology and Community-based Sustainable Development*, London: Belhaven.

Bhalla, S.S. and S. Bandyopadhyay, 1988. 'The Politics and Economics of Drought in India', paper presented at the Conference on Development Economics and Policy, Delhi School of Economics, Delhi, 18–21 December.

Bharara, L.P., 1992. 'Notes on the Experience of Drought: Perception, Recollection and Predictions' in B. Spooner and H.S. Mann (eds), *Desertification and Development: Dryland Ecology in Social Perspective*, New York: Academic Press.

——————, 1993. 'Indian Pastoral Communities Following Livestock Migration as a Drought Adjustment Strategy', *Journal of Anthropological Survey of India*, 42: 30–54.

Bhatia, B.M., 1989. 'Drought in Retrospect: Some Policy Issues', paper presented at the National Workshop on Management of Droughts, 4–5 July, New Delhi. (Published in GOI, 1989.)

Bidinger, F.R. and C. Johansen (eds), 1988. *Drought Research Priorities for the Dryland Tropics*, Patancheru: ICRISAT.

Binswanger, H.P., 1986. 'Risk Aversion, Collateral Requirements and Markets for Credit and Insurance in Rural Areas' in P. Hazell, C. Pomareda and A. Valdes, Crop Insurance to Agriculture Development, op. cit.

——————, S.M. Virmani and J. Kapen, 1980. *Farming Systems Components for Selected Areas in India: Evidence from ICRISAT, Research Bulletin No. 2*, Patancheru: ICRISAT.

Bjonness, I.M., 1983. 'External Economic Dependency and Changing Human Adjustment to Marginal Environments in High Himalaya, Nepal', *Mountain Research and Development* 3(3).

Blaikie, P.M., J. Harriss and A. Pain, 1985. 'The Management and Use of Common Property Resources in Tamil Nadu' in *Proceedings of the Conference on Common Property Resource Management*, Washington, DC: National Academy of Sciences.

—————, and H. Brookfield (eds), 1987. *Land Degradation and Society*, London: Methuen.

Blair, H., 1986. 'Social Forestry: Time to Modify Goals?' *Economic and Political Weekly* 21(30).

Borkar, V.V. and M.V. Nadkarni, 1975. *Impact of Drought on Rural Life*, Bombay: Popular Prakashan.

Brandon, K., 1995. 'People, Parks, Forests or Fields: A Realistic View of Tropical Forest Conservation', *Land Use Policy* 22(1).

Brara, R., 1987. *Shifting Sands: A Study of Rights in Common Pastures*, Jaipur: Institute of Development Studies.

Bromley, D.W. and D. Chapagain, 1984. 'The Village Against the Centre: Resource Depletion in South Asia', *American Journal of Agricultural Economics* 66(5), December.

—————, and M.M. Cernea, 1989. *The Management of Common Property Natural Resources: Some Conceptual and Operational Fallacies, World Bank Discussion Paper 57*, Washington, DC: The World Bank.

CAZRI (Central Arid Zone Research Institute), 1965. 'Socio-economic Survey of Livestock Breeders in Anupgadh Pugal Regions of Western Rajasthan', December (mimeo).

Chambers, R., 1987. *Sustainable Rural Livelihood: A Strategy for People, Environment and Development. Discussion Paper 240*, Sussex: Institute of Development Studies.

—————, 1990. *Micro Environments Unobserved. Gate Keepers Series No. 22*, London: International Institute for Environment and Development.

—————, A. Pacey and L.A. Thrupp (eds), 1989b. *Farmer First: Farmer Innovation and Agricultural Research*, London: Intermediate Technology Publications.

—————, N.C. Saxena and T. Shah, 1989b. *To the Hands of the Poor: Water and Trees*, New Delhi: Oxford and IBH Publishing Co.

Chand, R. and T. Haque, 1997. 'Sustainability of Rice-Wheat Crop System in Indo-Gangetic Region 1997', *Economic and Political Weekly* 32(13): A-26–A-30.

Chaudhari, K.M. and M.T. Bapat, 1975. *A Study of Impact of Famine and Relief Measures in Gujarat and Rajasthan*, Vallabh Vidyanagar (Gujarat): Agro-economic Research Centre, Sardar Patel University.

Chen, M.A., 1988. 'Size, Status and Use of Common Property Resources: A Case Study of Dhevdholera Village in Ahmedabad District, Gujarat', paper presented at Women and Agriculture Seminar, Trivandrum: Centre for Development Studies.

—————, 1991. *Coping with Seasonality and Drought*, New Delhi: Sage Publications.

Chopra, K., G.K. Kadekodi and M.N. Murthy, 1990. *Participatory*

Development. An Approach to the Management of Common Property Resources, New Delhi: Sage Publications.

Chopra, R., 1989. 'Voluntary Organizations in Drought Management', paper presented at National Workshop on Management of Droughts, 4–5 July, New Delhi, (GOI 1989) op. cit.

Collier, G.A., 1990. *Seeking Food and Seeking Money: Changing Relations in Highland Mexico Community. Discussion Paper No. 11*, Geneva: United Nations Research Institute for Social Development (UNRISD).

Conway, G.R., 1985. 'Agricultural Ecology and Farming Systems Research', in J.V. Remenyi (ed.), *Agricultural Systems Research for Developing Countries*, Canberra: Australian Centre for International Agricultural Research.

Dahlberg, K.A., 1987. 'Redefining Development Priorities: Genetic Diversity and Agroeco development', *Conservation Biology* 1(4).

Daly, H.E. and J.B. Cobb, Jr., 1989. *For the Common Good: Redirecting the Economy Towards Community, the Environment and Sustainable Future*, London: Merlin Press.

Dandekar, K. and M. Sathe, 1980. 'Employment Guarantee Scheme and Food for Work Programme', *Economic and Political Weekly* 15(15): 707–13.

Dandekar, V.M., 1976. 'Crop Insurance in India', *Economic and Political Weekly* 11(26), (Review of Agriculture).

Dasgupta, M., 1987. 'Informal Security Systems and Population Retention in Rural India', *Economic Development and Cultural Change* 35(2).

Dasgupta, P., 1997. *Environmental and Resource Economics in the World of the Poor*, Washington, DC: Resources for the Future.

DESFIL (Development Strategies for Fragile Lands), 1988. *Development of Fragile Lands, Theory and Practice*, Washington, DC: DESFIL.

Dev, S.M. and M.H. Suryanarayana, 1997. 'Is PDS Urban Biased or Pro-Rich?: An Evaluation', *Economic and Political Weekly* 32(41): 2357–66.

Downing, T.E., M.J. Watts and H.G. Bohle, 1996. 'Climate Change and Food Insecurity; Towards a Sociology and Geography of Vulnerability' in Downing, T.E. (ed.), 1996. *Climate Change and World Food Security*, Berlin: Springer.

Dregne, H.E., 1983. *Desertification of Arid Lands: Advances in Arid Land Technology and Development*, vol. 3, New York: Haswood.

Dreze, Jean, 1990. 'Famine Prevention in India' in J. Dreze and A. Sen (eds), ch. 2.

———, and A. Sen, 1990a. *Hunger and Public Action*, Oxford: Clarendon Press.

Eckholm, E.P., 1975. The Deterioration of Mountain Environments, *Science*, 139: 764–74.

El-Swaify, S.A., P. Pathak, T.J. Rego and S. Singh, 1985. 'Soil Management for Optimized Productivity under Rainfed Conditions in the Semi-arid Tropics', *Advances in Soil Science* 1, 1–647.

Economic and Political Weekly, 1973a. Rajasthan: 'Developing the Desert', *EPW* 8(25).

————, 1975c. Bombay: 'Keeping the Poor Out', *EPW* 10(13).

————, 1975a. Tamil Nadu: 'Starvation Deaths in a Surplus State', *EPW* 10(8).

————, 1975c. 'Drought: Many Uses', *EPW* 10(19).

————, 1972c. 'Field Report on Drought in Maharastra: For Most, Metal-Breaking', *EPW* 7(52).

————, 1972b. 'Parliament and Famine: Debating over the Dead', *EPW* 7(49).

————, 1970. 'The Drought: Hunger and Glory', *EPW* 5(1).

————, 1972d. 'The Drought: Unwanted Men', *EPW* 7(52).

————, 1975d. 'The Famine: Productivity Norms for Starving', *EPW* 10(41).

————, 1973b. 'West Bengal: The Hungry Flock into Calcutta', *EPW* 8(35).

————, 1972a. 'Jaisalmer: Developing Desert', 7(41).

————, 1971. Andhra: 'Starvation amidst Plenty', *EPW* 6(48).

————, 1985. 'Orissa Drought and Poverty: A Report from Kalahandi', *EPW* 20(2) (November 1985: 1857–60).

Farrington, J. and S.B. Mathema, 1991. *Managing Agricultural Research in Fragile Environments: Amazon and Himalayan Case Studies*, London: ODI.

Feeny, D.F., Berkes, B.J. McCay and J.M. Acheson 1990. 'The Tragedy of Commons: Twenty-two Years Later', *Human Ecology* 19(1).

Fisher, R.J., 1997. *If Rain Doesn't Come: An Anthropological Study of Drought and Human Ecology in Western Rajasthan*, New Delhi: Manohar Publishers.

Franke, R.W. and B.H. Chasin, 1980. *Seeds of Famine*, Ottawa: Allenheld, Osmum.

Gadgil, M., 1985. 'Towards and Ecological History of India', *Economic and Political Weekly* 20(45–47), Special Number.

————, and P. Iyer, 1989. 'On the Diversification of Common Property Resource User of by the Indian Society' in F. Berks (ed.), *Common Property Resources*.

————, and R. Guha, 1992. *This Fissured Land: An Ecological History of India*, Berkeley: University of California Press.

————, 1995. *Ecology and Equity: The Use and Abuse of Nature in Contemporary*, London: Routledge.

Gadgil, M., F. Berkes and C. Folke, 1993. *Indigenous Knowledge for Biodiversity Conservation'. Beijer Reprint Series, no. 15,* Stockholm: Beijer International Institute of Ecological Economics.

————, Gadgil, S., A.K.S. Huda, N.S. Jodha, R.P. Singh and S.M. Virmani, 1988. 'The Effects of Climatic Variations on Agriculture in Dry Tropical Regions of India' in M.L. Parry, T.C. Carter and N.T. Konijn (eds), *The Impact of Climatic Variation on Agriculture, vol. II, Assessment in Semi-Arid Regions*, London: Kluwer Academic Publishers.

Glantz, M.H. (ed.), 1987. *Desertification: Environment Degradation in and Around Arid Lands*, Boulder, Colo: Westview Press.

GOI, 1989. *The Drought of 1987: Responses and Management, vol. I, National*

Efforts; vol. II State Experiences, Proceedings of the National Workshop on Management of Droughts, New Delhi: Department of Agriculture and Cooperation, Ministry of Agriculture, Government of India.

—————, 1998. *Proceedings of the National Workshop on Watershed Approach for Managing Degraded Lands in India: Challenge for the 21st Century* (27–29 April 1998), New Delhi: Department of Wasteland Development, Ministry of Rural Areas and Employment, Government of India and Department for International Development (DIID) of the United Kingdom.

—————, 1990. *National Watershed Development Project for Rainfed Areas (NWDPRA): Guidelines,* Ministry of Agriculture, Department of Agriculture and Cooperation, New Delhi: Government of India.

Grainger, A., 1982. *Desertification: How People Can Make Deserts: How People Can Stop and Why They Don't,* London: Earthscan.

Guha, R., 1997. 'Socio-ecological Research in India—A Status Report', *Economic and Political Weekly* 32(7): 345–352.

Guillet, D.G., 1983. 'Toward a Cultural Ecology of Mountains: The Central Andes and the Himalayas Compared', *Current Anthropology* 24: 561–74.

Gupta, A.K. 1986. 'Socio-ecology of Stress: Why do Common Property Resource Management Projects Fail?' in *Proceedings of the Conference on Common Property Resource Management,* Washington, DC: BOSTID, National Research Council.

—————, 1991. 'Sustainability Through Biodiversity: Designing Crucible of Culture, Creativity and Conscience', paper presented at the International Conference on Biodiversity and Conservation at Danish Parliament, Copenhagen, 8 November.

Hazell, P., C. Pomareda and A. Valdes, 1986. *Crop Insurance for Agricultural Development: Issues and Experiences,* Baltimore: The Johns Hopkins University Press.

Hazra, C.R., D.P. Singh and R.N. Kaul, 1996. *Greeting Common Lands in Jhansi through Village Resource Development: A Case Study,* New Delhi: Society for Wasteland Development.

Hussain, S.S. and O. Erenstein, 1992. *Monitoring Sustainability Issues in Agricultural Development: A Case Study in Swat in North Pakistan, PATA Working Paper 6,* Saidu Sharif (NWFP, Pakistan), PATA, Integrated Agricultural Development Project.

ICRISAT, 1987. *Proceedings of the Workshop on Farming Systems Research, (17–21 February 1986),* Patancheru: International Crops Research Centre for Semi-Arid Tropics.

ICIMOD, 1995. *Community Forestry: The Language of Life (Report of the First Regional Community Forestry User's Group Workshop),* Kathmandu: ICIMOD.

IIED, 1995. *The Hidden Harvest. The Value of Wild Resources in Agricultural Systems (A Summary), Sustainable Agriculture Programme,* London: International Institute of Environment and Development (IIED).

Indrakanth, S., 1997. 'Coverage and Leakages in PDS in Andhra Pradesh',

Economic and Political Weekly 32(19): 999–1001.

Ives, J.D. and B. Messerli, 1989. *The Himalayan Dilemma: Reconciling Development and Conservation*, London: Routledge.

Iyengar, S., 1988. *Common Property Land Resources in Gujarat, Some Findings About Their Size, Status and Use. Working Paper 18*, Gota, Ahmedabad: The Gujarat Institute of Area Planning.

Jochim, M.A., 1981. *Strategies for Survival: Cultural Behaviour in an Ecological Context*, New York: Academic Press.

Jodha, N.S., 1968. 'Capital Formation in Arid Agriculture: A Study of Soil Conservation Measures Applied to Agriculture', PhD Thesis, University of Jodhpur, Jodhpur.

—————, 1969. 'Drought and Scarcity in Rajasthan Desert: Some Basic Issues', *Economic and Political Weekly* 4(16): 699–704.

—————, 1970. 'Land Policies of Rajasthan: Some Neglect Aspects', *Economic and Political Weekly* (26), Review of Agriculture, June.

—————, 1972a. 'Agricultural Development and Problems of Nomadic Tribals in the Rajasthan Desert' in M.L. Patel (ed.), *Agro-economic Problems of Tribal India*, Bhopal: Progress Publishers.

—————, 1972c. 'A Strategy for Dryland Agriculture', *Economic and Political Weekly*, (13) (Review of Agricultue).

—————, 1974. 'A Case of the Process of Tractorisation', *Economic and Political Weekly* 9(25): 111–118.

—————, 1975. ' Famine and Famine Policies: Some Empirical Evidence', *Economic and Political Weekly* 10(41): 16–23.

—————, 1978. 'Effectiveness of Farmers' Adjustment to Risk', *Economic and Political Weekly* 13(1): 38–48 (Quarterly Review of Agriculture).

—————, 1979. 'Dry Farming Technology: Achievements and Obstacles' in C.H. Shah and C.N. Vakil (eds), *Agriculture Development in India: Policy and Problems*, Bombay: Orient Longman.

—————, 1980. 'Intercropping in Traditional Farming Systems', *Journal of Development Studies* 16(4): 427–42.

—————, 1981. 'Role of Credit in Farmers' Adjustments Against Risk in Arid and Semi-arid Areas of India', *Economic and Political Weekly* 16 (22–23): 1696–1709.

—————, 1985a. 'Market Forces and Erosion of Common Property Resources' in *Proceedings of the International Workshop on Agricultural Markets in the Semi Arid Tropics*, 24–28 October , 1983, ICRISAT Centre, Patancheru, pp. 233–77.

—————, 1985b. 'Population Growth and the Decline of Common Property Resources in Rajasthan, India', *Population and Development Review* 11(2): 247–64.

—————, 1986a. 'Common Property Resources and Rural Poor in Dry Regions of India', *Economic and Political Weekly* 21(26): 1169–81.

—————, 1986b. 'Research and Technology for Dry Farming in India: Some Issues for the Future Strategy', *Indian Journal of Agricultural Economics* 41(3).

—————, 1988. 'Population Growth and Common Property Resources: Micro-level Evidence from India', *Proceedings of a United Nations Expert Group Meeting on Consequences of Rapid Population Growth in Less Developed Countries*, 23–26 August, Population Division, IESA, United Nations, 23–26 August, New York.

—————, 1989a. 'Dry Farming Research: Issues and Approaches' in N.S. Jodha (ed.), *Technology Options and Economic Policy for Dryland Agriculture*, Bombay: The Indian Society of Agricultural Economics, pp. 187–218.

—————, (ed.), 1989b. *Technology Options and Economic Policy for Dryland Agriculture: Potentials and Challenge*, Bombay: The Indian Society of Agricultural Economics (Concept Publishing Co.).

Jodha, N.S. 1989c. 'Potential Strategies for Adapting to Greenhouse Warming: Perspectives from the Developing World' in N.J. Robsenberg, W.E. Easterling and P.R. Crossman (eds) *Greenhouse Warming: Abatement and Adaptation*, Washington, DC: Resources for the Future.

—————, 1990a. 'Depletion of Common Property Resources' India: Micro-level Evidence' in G. McNicoll and M. Cain (eds) *Rural Development and Population: Institutions and Policy*, New York: Oxford University Press (also supplement to vol. 15, *Population and Development Review*).

—————, 1990b. 'Mountain Agriculture: The Search for Sustainability', *Journal of Farming Systems Research Extension* 1(1): 55–75.

—————, 1991a. 'Sustainable Agriculture in Fragile Resource Zones: Technological Imperatives', *Economic and Political Weekly* 26(13) (Review of Agriculture).

—————, 1991b. 'Agricultural Growth and Sustainability: Perspective and Experiences from the Himalayas' in S.A. Vosti, T. Reardon, von Uruff and J. Witcover, *Agricultural Growth and Poverty Alleviation: Issues and Policies*, Proceedings of the Conference, September 23–27, Food and Agriculture Development Centre, Feldafing, Germany. Also in Vosti et al. (eds), 1997 op. cit.

—————, 1992. *Rural Common Property Resources: A Missing Dimension of Development Strategies*, *World Bank Discusson Paper No. 169*. Washington, D.C.: World Bank.

—————, 1993. 'Understanding and Responding to Global Climate Change in Fragile Resource Zones' in J. Schmandt and J. Clarkson (eds), *The Regions and Global Warming: Impacts and Response Strategies*, New York: Oxford University Press.

—————, 1995a. *Sustainable Development in Fragile Environments: An Operational Framework for Arid, Semi-Arid and Mountain Areas*, Ahmedabad: Centre for Environmental Education.

—————, 1995b. *Field Notes on People Managed Biodiversity in India* (Limited Circulation). ENVSP, Environment Department, Washington, DC: The World Bank.

—————, 1996a. 'Ride the Crest or Resist the Change? Responses to Emerging Trends in Rainfed Farming Research in India', *Economic and Political Weekly* 31(28).

————, 1996b. 'Property Rights and Development' in S.S Hanna, C. Folke, K-G Maler (eds), *Right to Nature: Ecological, Economic, Cultural and Political Principles of Institutions for the Environment*, Washington, DC: Island Press.

————, 1997. 'Mountain Agriculture' in Messerli, B. and J.D. Ives (eds), *Mountains of the World: A Global Priority*, New York: The Parthenon Publishing Group.

————, 1999. 'Process of Rapid Globalisation and Repercussion's for Mountain Areas and Communities', paper under review for publication in *Mountain Research and Development*.

————, and A.C. Mascarenhas, 1985. 'Adjustment in Self-Provisioning Societies, in R.W. Kates, J.H. Ausubel and M. Berberian (eds), *Climate Impact Assessment: Studies of the Interaction of Climate and Society*, London: John Wiley and Sons, pp. 437–64.

————, M. Banskota and T. Partap (eds), 1992. *Sustainable Mountain Agriculture* (2 vols), New Delhi: Oxford and IBH Publishing Co. Pvt. Ltd.

————, and R.P. Singh, 1982. 'Factors Constraining Growth of Coarse Grain Crops in Semi-Arid Tropical India', *Indian Journal of Agricultural Economics* 37(3): 346–54.

————, 1990. 'Crop Rotations in the Traditional Farming System in Selected Areas of India', *Economic and Political Weekly* (Review of Agriculture) vol. 25 (13): A 28-A 35.

————, and T. Partap, 1993. 'Folk Agronomy in the Himalayas: Implications for Agricultural Research and Technology' in *Rural People's Knowledge, Agricultural Research and Extension Practices*, IIED Research Series, 3(1), IIED, London.

————, and V.S. Vyas, 1969. *Conditions of Stability and Growth in Arid Agriculture*, Vallabh Vidyanagar: Agro-Economic Research Centre.

Karanth, G.K., 1991. 'Farmers' Survival Strategies in a Drought-Prone District—A Case Study of a Village in Chitradurga District' (mimeo), Bangalore: ADRT Unit, Institute for Social and Economic Change.

Kasperson, R.E., K. Dow, D. Golding and J.X. Kasperson (eds), 1990. *Understanding Global Environmental Change: The Contributions of Risk Analysis and Management*, Clark University, Worcester.

Kates, R.W. and S. Millman, 1990. 'On Ending Humger: The Lessons of History in L.F. Newman et al. (eds), *Hunger in History. Food Shortage, Poverty and Deprivation*', Oxford: Basil Blackwell.

Kaul, Minoti. 1987. *Common Land in Delhi: The Bisgama Cluster Report*, New Delhi.

Kavoori, P.S., 1999. *Pastoralism in Expansion: Transhuming Sheep Herders of Western Rajasthan*, New Delhi: Oxford University Press.

Kerr, J.M., N.K. Sanghi and Srivaramappa, 1996. *Subsidies in Watershed Development Projects in India: Distortions and Opportunities*, IIED Gate Keeper Series No. 61, London: International Institute for Environment and Development.

Kothari, A., 1997. *Conserving India's Agro-Biodiveristy: Prospects and Policy Implications*, Gate Keeper Series No. 65, London: International Institute of Environment and Development (IIED).

———— and R.V. Anuradha, 1997. 'Biodiversity, Intellectual Property Rights and GATT Agreement: How to Address the Conflict', *Economic and Political Weekly* 32(43): 2814–28.

Krishna Kumar, A., 1994. 'Harnessing a Heritage: The Rights of Local Communities', *Frontline* 11 March.

Krishna, A., N. Uphoff and M.J. Esman, 1997. *Reasons for Hope: Instructive Experiences in Rural Development*, West Hartford, USA: Kumarian Press.

Krishna, S., 1996. 'The Environmental Discourse in India: New Directions in Development' in T.V. Sathyamurthy (ed.), *Social Change and Political Discourse in India*, vol. 4, New Delhi: Oxford University Press.

Kumar, P., V. Lobu, J. Chatterji and A. Khare, 1992. 'Common Property and Local Institutions', *Wasteland News* 7(2): 3–10.

Ladejinsky, W., 1972. 'Land Ceilings and Land Reforms', *Economic and Political Weekly* 7 (5–7).

Leach, M., R. Mearn and I. Scoones, 1997. *Environmental Entitlements: A Framework for Understanding Institutional Dynamics of Environmental Change. IDS Discussion Paper 359*, Brighton: Institute of Development Studies.

Limaye, B.V. and M.D. Rahalkar, 1971. 'Drought Relief: A First Hand Account', *Economic and Political Weekly* (41).

Lynam, J.K. and R.W. Herdt, 1988. 'Sense and Sustainability: Sustainability as an Objective in International Agriculture Research', paper presented to CIP–Rockefeller Foundation Conference on 'Farmers and Food Systems', Lima, Peru, 26–30 September.

Mailik, A.K. and T.S. Govindaswamy, 1962–63. 'The Drought Problem of India in Relation to Agriculture', *Annals of Arid Zone* 1 (1 and 2).

Mann, H.S. and S.K. Saxena, 1980. *Khejri (Proposes cineraria) in the Indian Desert: Its Role in Agroforestry. CAZRI Monograph, no. 11*, Jodhpur: CAZRI.

Markandya, A. and D.W. Pearce, 1988. *Environmental Considerations and the Choice of the Discount Rate in Developing Countries, Environment Department Working, Paper No. 3*, Washington, DC: The World Bank.

Mathur, K., 1989. 'Politics and Management of Drought in India', paper presented at the National Workshop on Management of Droughts, 4–5 July, New Delhi. (GOI 1989, op. cit.)

Meti, T.K., 1976. 'Economy of Scarcity Areas', Dharwar: Karnataka University (Economics Series No. 4).

Metz, J.J., 1991. 'A Reassessment of Causes and Severity of Nepal's Environmental Crisis', *World Development* 19(7).

Miller, K.R., 1995. 'Balancing the Scales: Policies for Increasing Biodiversity's Chances through Bioregional Management' (Draft). Washington, DC: World Resource Institute.

Mishra, P.R. and M. Sarin, 1987. 'Sukhomajri-Nada: A New Model of Eco-Development', *Business India* (Bombay), 16–29 November.

Mody, Navroz, 1972. 'Field Report on Drought in Maharashtra: to Some a God Send', *Economic and Political Weekly* 7(52).

Morris, David Morris, 1974. 'What is Famine', *Economic and Political Weekly* 9(44).

——————, 1975. 'Needed: A New Famine Policy', *Economic and Political Weekly* (Annual Number) 1975.

Mulk, M. 1992. *The Farmer's Strategies for Sustainable Mountain Agriculture: Chitral District, Pakistan*, MFS Discussion Paper 28. ICIMOD, Kathmandu.

Munasinghe, M., 1993. *Environmental Economics and Sustainable Development, The World Bank Environment Paper No.3*, Washington, DC: The World Bank.

Nadkarni, M.V., 1985. *Socio-Economic Conditions in Drought-Prone Areas*, New Delhi: Concept Publishing Company.

——————, S.A. Pasha and L.S. Prabhakar, 1989. *The Political Economy of Forest Use and Management*, New Delhi: Sage Publications.

Nagaraj, N. and M.G. Chandrakanth, 1997. 'Intra- and Inter-Generational Equity Effects of Irrigation Well Failures: Farmers in Hard Rock Areas of India', *Economic and Political Weekly* 32(13) (Review of Agriculture).

Nair, P.K., 1983. 'Tree Integration on Farming Lands for Sustained Productivity of Small Holdings' in W. Lockertz (ed.), *Environmentally Sound Agriculture*, New York: Praeger.

Nelson, R., 1988. *Dryland Management: The Desertification Problem, Environment Department Working Paper No. 8*, Washington, DC: The World Bank.

Norgaard, R.B., 1984. 'Coevolutionary Agricultural Development', *Economic Development and Cultural Change*, 60: 160–73.

——————, 1999. 'Beyond Growth and Globalisation', *Economic and Political Weekly* 34(36).

ODA, 1989. *Evaluation of the Social Forestry Project in Karnataka, India*, London: Overseas Development Administration.

O'Riordan, T., 1988. 'The Politics of Sustainability' in R.K. Turner (ed.), *Sustainable Environment Management: Principle and Practice*, London: Belhaven Press.

Ostrom E., 1988. 'Institutional Arrangements and the Commons Dilemma' in E. Ostrom, D. Feeny and H. Picht (eds), *Rethinking Institutional Analysis and Development*, International Centre for Economic Growth, Institute for Contemporary Studies, USA.

Oza, A., 1989. 'Availability of CPR Lands at Micro-Level: Case Studies of Junagad Programme Area of AKRSP (India)' in *Status of Common Property Land Resources in Gujarat and Problem of Their Development*, Ahmedabad: Gujarat Institute of Area Planning.

Panayotou, T., 1990. *The Economics of Environmental Degradation: Problems, Causes and Responses. Development Discussion Paper No. 335*, Cambridge, MA: HIID, Harvard University.

Pandey, S.M. and J.N. Upadhyay, 1976. *Effects of Drought on Rural Population: A Study in Jhajjar, Haryana*, New Delhi: Shri Ram Centre for Industrial Relations and Human Resources.

Pant, D.D., 1935. *The Social Economy of Himalayas: Based on a Survey in the Kumaon Himalayas*, London: George Allen and Unwin.

Parry, M.L., T.R. Carter and N.T. Konijin, 1988. *The Impacts of Climate Variations on Agriculture: Assessment in Semi-Arid Regions*, vol. 2, Dordrecht: Kluwer Academic Publishers.

Parthasarathy, G. and Shameen, 1998. 'Suicides of Cotton Farmers in Andhra Pradesh: An Exploratory Study', *Economic and Political Weekly* 33(13): 720–26.

Patil, S., 1973. 'Famine Conditions in Maharastra: A Survey of Sakari Taluka', *Economic and Political Weekly* 8(30).

Pezzey, J., 1989. *Economic Analysis of Sustainable Growth and Sustainable Development, Environment Department Working, Paper No. 15*, Washington, DC: The World Bank.

Poffenberger, M. and B. McGean, 1996. *Village Voices, Forest Choices: Indian Experiences in Joint Forest Management into the 21st Century*, New Delhi: Oxford University Press.

Prabhakar, M.S., 1974. 'The Famine: A Report from Dhubri', *Economic and Political Weekly* 9(42).

Pradhan, J., 1993. 'Drought in Kalahandi: The Real Story', *Economic and Political Weekly* 28(22): 1084–88.

Prakash, S., 1997. *Poverty and Environment Linkages in Mountains and Uplands: Reflections on the Poverty Thesis. CREED Working Paper Series No. 12*, London: International Institute of Environment and Development.

Prasad, N.P. and P. Venkatarao, 1996. 'Patterns of Migrations from a Drought-Prone Area Andhra Pradesh: Case Study', *Journal of Human Ecology* 7(2): 1–8.

Pretty, J., 1995. *Regeneration Agriculture: Politics and Practices for Sustainability and Self Reliance*, London: Earthscan.

Radhakkrishna, R. and C.H. Rao, 1994. *Food Security, Public Distribution System and Price Policy, Research Project on Strategies and Financing for Human Development in India*, UNDP: Thiruvananthapuram.

Raeburn, J.R., 1984. *Agriculture: Foundations, Principles and Development*, New York: John Wiley and Sons.

Rai, R., 1942. *Akal Kashta Niwarak* (in Hindi) (A Report on Famine-Scarcity Eradication to the Councilor of Jodhpur State), Bali, Marwar.

Rangaswami, Amrita, 1974. 'Financing Famine Relief: Calling a Bluff', *Economic and Political Weekly* 9(45, 46).

————, 1975. 'The Uses of Drought', *Economic and Political Weekly* 10(50).

————, 1989, 'Financing in the Management of Droughts: Policy Perspectives', paper presented at National Workshop on Management of Droughts, 4–5 July 1989, New Delhi.

Rao, C.H. Hanumantha, 1994. *Agricultural Growth, Rural Poverty and Environmental Degradation in India*, New Delhi: Oxford University Press.

Repetto, R. and T. Holmes, 1984. 'The Role of Population in Resource Depletion in Developing Countries', *Population and Development Review* 9(4).

Reynolds, N. and P. Sundar, 1977. 'Maharashtra's Employment Guarantee Scheme: A Programme to Emulate', *Economic and Political Weekly* 12(29).

Rieger, H.C., 1981. 'Man Versus Mountain: The Destruction of the Himayayan Ecosystem' in J.S Lall and A.D. Moodie (eds), *The Himalaya: Aspects of Change*, New Delhi: Oxford University Press.

Rongsen, Lu. 1998. *Enterprises in Mountain Specific Products in Western Sichuan, China*, MEI Discussion paper 98/7, Kathmandu: ICIMOD.

Roy, S., 1996. 'Development, Environment and Poverty, Some Issues for Discussion', *Economic and Political Weekly* 31 (27 January).

—————, 1997. Globalisation, Structural Change and Poverty: Some Conceptual and Policy Issues, *Economic and Political Weekly* 32(33–34).

Runge, C.F., 1981. 'Common Property Externalities: Isolation, Assurance and Resource in a Traditional Grazing Context', *American Journal of Agricultural Economics* 63(4).

Ruttan, V.W., 1988. 'Sustainability is Not Enough', paper presented at Symposium on Creating Sustainable Agriculture for the Future, University of Minnesota, St. Pauls, 30 April.

————— (ed.), 1989. *Biological and Technological Constraints on Crop and Animal Productivity: Report on a Dialogue, Staff Paper Series*, St. Paul: Department of Agricultural and Applied Economics, University of Minnesota.

Sanwal, M., 1989. 'What We know about Mountain Development: Common Property, Investment Priorities and Institutional Arrangements', *Mountain Research and Development* 9(1), Boulder, Colo.

Sarin, M., 1996. *Joint Forest Management: The Haryana Experience*, Ahmedabad: Centre for Environment Education.

Sen, A. 1981. *Poverty and Famines: An Essay on Entitlement and Deprivation*, Oxford: Oxford University Press.

Shah, A., 1998. 'Watershed Development Programmes in India: Emerging Issues for Environment Development Perspectives', *Economic and Political Weekly* 33(26) (Review of Agriculture).

Shah, T., 1987. 'Profile of Collective Action on Common Property: Community Fodder Farm in Kheda District', Anand: Institute of Rural Management.

Shankarnaryan, K.A. and J. Kalla, 1985. *Management Systems for Natural Vegetation*, Jodhpur: Central Arid Zone Research Institute.

Sharma, P and T. Partap 1993. *Population Poverty and Development Issues in the Hindu Kush-Himalayas*, paper presented at the International Forum on Development of Poor Mountain Area, March 22–27, Beijing, China.

Shiva, V. 1996. *Future of Our Seeds, Fututre of Our Farmers,* New Delhi: Research Foundation for Science, Technology and Natural Resource Policy.

Shrestha, S., 1992. *Mountain Agriculture: Indicators of Unsustainability and Options for Reversal,* MFS Discussion Paper 32. ICIMOD, Kathmandu.

Shutain, G. and H. Chunru, 1989. *Problems of the Environment in Chinese Agriculture and a Strategy for Ecological Development* (an overview), Ministry of Agriculture and Beijing Agricultural University, Beijing, China.

SIDA, 1988. *Forestry for the Poor: An Evaluation of the SIDA Supported Social Forestry Project in Tamil Nadu, India*, Stockholm: SIDA.

Singh, V., 1992. 'Dynamics of Unsustaibility of Mountain Agriculture' (An unpublished report on UP hill areas, India), ICIMOD, Kathmandu.

Singh, C., 1986. *Common Property and Common Poverty*, New Delhi: Oxford University Press.

Singh, K., 1995. *The Watershed Approach to Sustainability of Renewable Common Pool Natural Resources: Lessons from India's Experience. IRMA Research Paper No. 14*. Anand: Institute of Rural Management.

Singh, N.K., 1972. 'Bihar: Hundred Years After', *Economic and Political Weekly* 7(19).

Singh, V., 1995. 'Biodiversity and Farmers: Experiences from Garhwal Himalaya, India', paper presented at the Beijer Research Seminar at Kota Kinabalu, Malaysia, 16–19 May.

South Centre, 1996. 'Liberalisation and Globalisation: Drawing Conclusions for Development', Geneva: South Centre.

SPWD, 1991. 'National Policy on Common Property Land: Draft Paper', *Wasteland News* 6(4), New Delhi: Society for Wasteland Development.

Subramanian, V., 1975. *Parched Earth: The Maharashtra Drought 1972–73*, Bombay: Orient Longmans Limited.

Swaminathan, M.C., K.V. Rao and H. Rao, 1969. 'Food and Nutrition Situation in the Drought Affected Area of Bihar', *Journal of Nutrition and Diet* (6), 209–17.

Swarup, R., 1991. *Agricultural Economy of Himalayan Region*, G. B. Pant Institute of Himalayan Environment and Development, Almora, India.

Tisdell, C., 1988. 'Sustainable Development: Differing Perspectives of Ecologists Economists, and Relevance to Less Developed Countries', *World Development* 16(3).

Topping, J.C., A. Qureshi and S.A. Sherer, 1991. 'Implications of Climate Change for the Asian and Pacific Region', paper presented at the Asian Pacific Seminar on Climate Change, 23–26 January 1991. Nagoya, Japan.

Turner, B.L., R.E. Kasperson, W.B. Meyer, K. M. Dow, Dolding, J.X. Kasperson, R.C. Mitchell and S.J. Ratick, 1990. Two Types of Global Environmental Change: Definitional and Spatial-scale Issues in Their Human Dimension, *Global Environment Change* 1(1): 14–22.

UNDP, 1999. *Human Development Report 1999*, New York: Oxford University Press.

Venugopal, K.R., 1992. *Deliverance from Hunger: The Public Distribution System in India*, New Delhi: Sage Publications.

Virmani, S.M., M.V.K. Shiva Kumar and R.P. Sarkar, 1982. 'Rainfall Probability and Tailoring Agriculture to Match It' in *Symposium on Rainwater and Dryland Agriculture*, 3 October 1980, New Delhi: Indian National Science Academy.

Vosti, S.A. and T. Reardon (eds), 1997. *Sustainability, Growth and Poverty Alleviation: A Policy and Agro-ecological Perspective*, Baltimore: The Johns Hopkins University Press.

Vyas, V.S. and V.R. Reddy, 1998. 'Assessment of Environmental Policies and Policy Implementation in India', *Economic and Political Weekly* 33 (1–2): 48–55.

Wade, R., 1988. *Village Republics: Economic Conditions for Collective Action in South India*, Cambridge: Cambridge University Press.

Walker, T.S. and N.S. Jodha, 1986. 'How Small Farm Households Adapt to Risk' in P. Hazell et al. (eds), op. cit.

Walker, T.S. and J.G. Ryan, 1990. *Village and Household Economics in India's Semi-Arid Tropics*, Baltimore: The Johns Hopkins University Press.

Warren, A. and C. Agnew, 1988. *An Assessment of Desertification and Land Degradation in Arid and Semi-arid Areas, Dryland Programme Document, no. 2*, London: International Institute for Environment and Development.

Whitaker, Meri L., P.V. Shenoi and J.M. Kerr, 1991. 'Agricultural Sustainability, Growth and Poverty Alleviation in the Indian Semi-Arid Tropics' in S.A. Vosti, Reardon, W. von Urff and J. Witcover (eds), *Agricultural Sustainability, Growth and Poverty Alleviation: Issues and Policies*, Proceedings of the conference, 23–27 September, Food and Agriculture Development Centre, Feldafing, Germany. Also in S.A. Vosti and T. Reardon (eds), 1994, 1997. *Sustainability, Growth and Poverty Alleviation: A Policy and Agro-ecological Perspective*, Baltimore: Johns Hopkins University Press.

Whiteman, P.T.S., 1988. 'Mountain Agronomy in Ethiopia, Nepal and Pakistan' in N.J.R Allan et al. (eds), op. cit.

Wolf, Ladejinsky, 1973. 'Drought in Maharastra: Not in Hundred Years', *Economic and Political Weekly* 8(7).

World Bank, 1989. *Asia Region Review of Watershed Development Strategies and Technologies (A Report of the Asia Region Technical Department, Agriculture Division and Environment Department, Policy and Research Division)*, Washington, DC: World Bank.

World Bank, 1999. *Entering the 21st Century—World Development Report 1999–2000*, New York: Oxford University Press.

Yadav, Y., 1992. 'Farming-Forestry-Livestock Linkages: A Component of Mountain Farmers' Strategies' in N.S. Jodha, M. Banskota and T. Partap (eds), *Sustainable Mountain Agriculture*, New Delhi: Oxford and IBH Publishing.

Zazueta, A., 1995. *Policy Hits the Ground: Participation and Equity in Environmental Policy Making*, Washington, DC: World Resource Institute.

Zerner, C., 1999. *Justice and Conservation Insights from People, Plants and Justice: The Politics of Nature Conservation*, New York: The Rain Forest Alliance.

Index